Solutions Manual : Fundamentals of Communication Systems

Janak Sodha

Janak Sodha received his B.Sc. (Hons.), M.Sc. and Ph.D. from the University of Manchester, England (UK). A telecommunications research engineer with ERA Technology (UK) for two years and currently a senior lecturer at the University of the West Indies, Cave Hill Campus, Barbados. This book is based on the experience of lecturing on communication theory for over eighteen years. Refer to the author's home page on http://janaksodha.com for details of his publications and search for the "Communications Systems" app for your smartphone or tablet using the author's surname.

Solutions Manual : Fundamentals of Communication Systems
Copyright © 2015 AppBooke Publishing (UK)

ISBN 978-0-9928510-1-9

Contents

Preface

This is the solutions manual for the text "**Fundamentals of Communication Systems**", **ISBN 9780992851002** which contains the following chapters:

- Signals

- Analog Communications

- Digital Communications

- Information Theory

- Analog to Digital

- Baseband Signalling

- Bandpass Signalling

- Block and Convolutional Codes

In addition to full written solutions, you will find video solutions for select questions, in which the author solves the problem for you. The aim being to **experience the thinking process**. Furthermore, you will find simulation solutions, which provide you with the wonderful opportunity of **experimenting with the variables of the questions**. For academic staff, you may use the simulation solution to create a new problem.

This icon indicates that a **video solution** for the question is available via the **Communication Systems** app.

This icon indicates that a simulation solution (MATLAB and Mathcad) for the question is available via the **Communication Systems** app.

All the multimedia referenced in this book and the main text are delivered via the app in the iOS and Android app stores. Together with the source code, PDFs of all the simulations with results are made available to help students easily follow the simulation code.

Chapter 1

Signals

1.1 Solutions

1. (a)   The average voltage $\langle v(t) \rangle$ across the resistor is given by

$$\langle v(t) \rangle \;=\; \frac{1}{T} \int_{-T/2}^{T/2} v(t)\,dt \tag{1.1}$$

$$=\; \frac{1}{T} \int_{-T/2}^{T/2} V_o \cos\left(2\pi f t\right) dt = \frac{V_o}{T}\left[\frac{\sin\left(2\pi f t\right)}{2\pi f}\right]_{-T/2}^{T/2}$$

Since $f = \frac{1}{T}$ and $\sin(\pi) = 0$, we find $\langle v(t) \rangle = 0$ as expected. The instantaneous power $p(t)$ delivered to the resistor is given by $p(t) = i^2(t)R = \frac{v^2(t)}{R}$ where $i(t)$ is the instantaneous current through the resistor. By normalizing the resistance to $1\ \Omega$, the average power $\langle p(t) \rangle$ is given by

$$\langle p(t) \rangle = \langle i^2(t) \rangle = \langle v^2(t) \rangle = \frac{1}{T} \int_{-T/2}^{T/2} V_o^2 \cos^2\left(2\pi f t\right) dt \tag{1.2}$$

Making use of the trigonometry identity $\cos\left(2\theta\right) = 2\cos^2\theta - 1$

$$\langle p(t) \rangle = \frac{V_o^2}{2T} \int\limits_{-T/2}^{T/2} (1 + \cos(4\pi f t)) \, dt \tag{1.3}$$

$$= \frac{V_o^2}{2T} \left[t + \frac{\sin(4\pi f t)}{4\pi f} \right]_{-T/2}^{T/2} = \frac{V_o^2}{2} \tag{1.4}$$

(b) The signal has a finite duration and is therefore an energy signal. Signal energy

$$E = \int\limits_{-\infty}^{\infty} s(t)^2 dt = \int\limits_{-1}^{1} 4^2 dt = 16 \left[t \right]_{-1}^{1} = 16 \left(1 - (-1) \right) = 32 \text{ Joules} \tag{1.5}$$

- ⭐ SIMULATION **P1:** Shown that the signal power goes to zero as $T \to \infty$. Repeat this problem for the other signals presented in this simulation for practice.

2. The signal

$$s(t) = 2 \cos \left(5\pi t - \frac{2\pi}{3} \right) + 3 \cos \left(10\pi t - \frac{\pi}{2} \right) \tag{1.6}$$

$$= 2 * 1 \cos \left(2\pi 2.5 t - \frac{120\pi}{180} \right) + 2 * 1.5 \cos \left(2\pi 5 t - \frac{90\pi}{180} \right) \tag{1.7}$$

from which we can extract the amplitudes, frequency and phase.

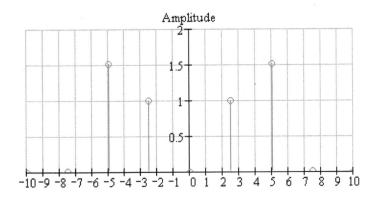

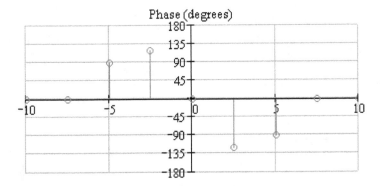

- SIMULATION **P2:** For the full solution.

- Experiment with the variables to understand and appreciate their influence on the answer.

3. Given that

$$c_n = \frac{1}{T} \int_T s(t) e^{\frac{-j2\pi nt}{T}} \, dt \qquad (1.8)$$

we insert $s(t)$ and integrate over the range 0 to τ, so that

$$c_n = \frac{1}{T}\int_0^\tau Ae^{\frac{-j2\pi nt}{T}}\,dt = \frac{A}{T}\left[\frac{-Te^{\frac{-j2\pi nt}{T}}}{j2\pi n}\right]_0^\tau \tag{1.9}$$

$$= \frac{A}{j2\pi n}\left[-e^{\frac{-j2\pi n\tau}{T}} - (-1)\right] \tag{1.10}$$

$$= \frac{A}{j2\pi n}\left[1 - e^{\frac{-j2\pi n\tau}{T}}\right] \tag{1.11}$$

$$= \frac{A}{j2\pi n}e^{\frac{-j2\pi n\tau}{2T}}\left(e^{\frac{j2\pi n\tau}{2T}} - e^{\frac{-j2\pi n\tau}{2T}}\right) \tag{1.12}$$

(note that $e^{-j\frac{\theta}{2}}\left(e^{-j\frac{\theta}{2}}\right) = e^{-j\left(\frac{\theta}{2}+\frac{\theta}{2}\right)} = e^{-j(\theta)}$)

But since $\dfrac{e^{j\theta} - e^{-j\theta}}{2j} = \sin\theta,$

$$c_n = \frac{A}{\pi n}e^{\frac{-j2\pi n\tau}{2T}}\sin\left(\frac{\pi n\tau}{T}\right) \tag{1.13}$$

$$= \frac{\tau}{T}\frac{AT}{\pi n\tau}e^{\frac{-j2\pi n\tau}{2T}}\sin\left(\frac{\pi n\tau}{T}\right) \tag{1.14}$$

Therefore,

$$c_n = \frac{A\tau}{T}e^{\frac{-j2\pi n\tau}{2T}}\frac{\sin\left(\frac{\pi n\tau}{T}\right)}{\frac{\pi n\tau}{T}} = \frac{A\tau}{T}\operatorname{sinc}\left(\frac{n\tau}{T}\right)e^{\frac{-j\pi n\tau}{T}} \tag{1.15}$$

• ⭐ SIMULATION **P3:** For part (b).

4. ★ SIMULATION **P4:** For the full solution. Experiment with the variables to understand and appreciate their influence on the answer.

5. ★ SIMULATION **P5:** For the full solution. Experiment with the variables to understand and appreciate their influence on the answer. Note that using this Mathcad/MATLAB code, you can solve any given Fourier exponential series problem.

6. ★ SIMULATION **P6:** For the full solution. Experiment with the sinusoidal frequency and observe how the spectrum is shifted.

7. Over a period $T = 2$ seconds, $s(t) = \begin{cases} t+1 & -1 \le t \le 0 \\ -t+1 & 0 \le t \le 1 \end{cases}$. Therefore

$$c_n = \frac{1}{T} \left[\int_{-T/2}^{0} (t+1)e^{\frac{-j2\pi nt}{T}} \, dt + \int_{0}^{T/2} (-t+1)e^{\frac{-j2\pi nt}{T}} \, dt \right] \qquad (1.16)$$

Have some fun simplifying this expression for yourself :-). Its not easy but good for the soul.

- ★ SIMULATION **P7:** For the solution. Experiment with the triangular waveform to your hearts content.

8. Let the rectangular pulse shown below be represented by

$$\text{rect}\left(\frac{t}{\tau}\right) = \begin{cases} 1 & -\frac{\tau}{2} \le t \le \frac{\tau}{2} \\ 0 & otherwise \end{cases} \qquad (1.17)$$

where $\tau = 2$.

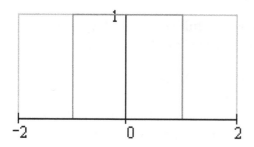

Recall that $A\operatorname{rect}\left(\dfrac{t}{\tau}\right) \leftrightarrow A\tau\operatorname{sinc}(f\tau)$. Using the symmetry property,

$$A\tau\operatorname{sinc}(t\tau) \leftrightarrow A\operatorname{rect}\left(\frac{f}{\tau}\right) \tag{1.18}$$

Thus for $s(t) = 2AB\operatorname{sinc}(2Bt)$, in which $\tau = 2B$, the solution is

$$A\operatorname{rect}\left(\frac{f}{2B}\right) \tag{1.19}$$

- ⭐ SIMULATION P8: Verification of the solution and for the answers to parts (b), (c) and (d). Experiment with the since pulse width and also with the low-pass filter e.g. set $B = \frac{1}{2}$ Hz and observe the fltered signal. Modify the code to determine the FT of various types of pulses.

9. For $s(t) = A\sin(2\pi f_c t)$, we make use of Eulers theorem, $\sin(2\pi f_c t) = \frac{1}{2j}\left(e^{j2\pi f_c t} - e^{-j2\pi f_c t}\right)$

$$S(f) = \int_{-\infty}^{\infty} \frac{A}{2j}\left(e^{j2\pi f_c t} - e^{-j2\pi f_c t}\right)e^{-j2\pi ft}\,dt \tag{1.20}$$

$$= \frac{A}{2j}\int_{-\infty}^{\infty}\left(e^{j2\pi f_c t}e^{-j2\pi ft} - e^{-j2\pi f_c t}e^{-j2\pi ft}\right)dt \tag{1.21}$$

$$= \frac{A}{2j} \int_{-\infty}^{\infty} \left(e^{j2\pi(f_c-f)t} - e^{-j2\pi(f_c+f)t} \right) dt \tag{1.22}$$

$$= \frac{A}{2j} \int_{-\infty}^{\infty} e^{-j2\pi(f-f_c)t} dt - \frac{A}{2j} \int_{-\infty}^{\infty} e^{-j2\pi(f+f_c)t} dt \tag{1.23}$$

$$= \frac{jA}{2} \int_{-\infty}^{\infty} e^{-j2\pi(f+f_c)t} dt - \frac{jA}{2} \int_{-\infty}^{\infty} e^{-j2\pi(f-f_c)t} dt \tag{1.24}$$

$$= j\frac{A}{2} \left[\delta\left(f + f_c \right) - \delta\left(f - f_c \right) \right] \tag{1.25}$$

where we made use of the Dirac property $\delta\left(f \right) = \int_{-\infty}^{\infty} e^{-j2\pi ft} dt$. Notice how this is simply the inverse FT of a d.c signal $s(t) = 1$.

• ⭐ SIMULATION **P9:** For the solution via Mathcad/MATLAB.

10. (a) For $s(t) = \begin{cases} e^{-\frac{t}{T}} & t > 0 \\ 0 & \text{otherwise} \end{cases}$, the FT is given by

$$S(f) = \int_{-\infty}^{\infty} s(t) e^{-j2\pi ft} dt = \int_{0}^{\infty} e^{-\frac{t}{T}} e^{-j2\pi ft} dt \tag{1.26}$$

$$= \int_{0}^{\infty} e^{-t(\frac{1}{T}+j2\pi f)} dt = \left[-\frac{e^{-t(\frac{1}{T}+j2\pi f)}}{\left(\frac{1}{T} + j2\pi f \right)} \right]_{0}^{\infty} = \frac{1}{\left(\frac{1}{T} + j2\pi f \right)} \tag{1.27}$$

$$= \frac{T}{(1 + j2\pi fT)}$$

● ⭐ SIMULATION **P10:** For the solution to parts (a), (b) and (c). Experiment with the low-pass filter and observe the corresponding filtered signal. Modify the signal expression to solve this problem for other types of pulses. If you take time to understand the simulation code, you will be able to easily modify my Mathcad/MATLAB files to solve a variety of other problems. *Honest labor never goes in vain !!!*

(d) If $s(t) = e^{-a|t|} = \begin{cases} e^{-at} & t > 0 \\ e^{at} & t < 0 \end{cases}$, then

$$S(f) = \int_{-\infty}^{\infty} s(t)e^{-j2\pi ft}dt = \int_{-\infty}^{0} e^{at}e^{-j2\pi ft}dt + \int_{0}^{\infty} e^{-at}e^{-j2\pi ft}dt \qquad (1.28)$$

$$= \int_{-\infty}^{0} e^{t(a-j2\pi f)}dt + e^{-t(a+j2\pi f)}dt \qquad (1.29)$$

$$= \left[\frac{e^{t(a-j2\pi f)}}{(a-j2\pi f)}\right]_{-\infty}^{0} + \left[-\frac{e^{t(a+j2\pi f)}}{(a+j2\pi f)}\right]_{0}^{\infty}$$

$$= \frac{1}{(a-j2\pi f)} + \frac{1}{(a+j2\pi f)} \qquad (1.30)$$

$$= \frac{2a}{(a-j2\pi f)(a+j2\pi f)} = \frac{2a}{a^2 + (2\pi f)^2}$$

11. The damped sinusoid pulse $x(t) = \begin{cases} e^{-t/T}\sin\left(\dfrac{2\pi t}{T}\right) & t > 0 \\ 0 & \text{otherwise} \end{cases}$ .Given

that the FT $S_{\text{mod}}(f)$ of the signal $s(t)\cos(2\pi f_c t + \theta)$ is given by

$$S_{\text{mod}}(f) = \frac{1}{2}\left[e^{j\theta}S(f - f_c) + e^{-j\theta}S(f + f_c)\right] \qquad (1.31)$$

where $s(t) \longleftrightarrow S(f)$ and

$$\cos(2\pi f_c t - \frac{\pi}{2}) = \cos(2\pi f_c t)\cos\left(\frac{\pi}{2}\right) + \sin(2\pi f_c t)\sin\left(\frac{\pi}{2}\right) = \sin(2\pi f_c t) \tag{1.32}$$

we have

$$e^{-t/T}\sin\left(\frac{2\pi t}{T}\right) = s(t)\cos(2\pi f_c t + \theta) \tag{1.33}$$

where

$$s(t) = e^{-t/T}, f_c = \frac{1}{T}, S(f) = \frac{T}{(1 + j2\pi fT)} \tag{1.34}$$

from the previous problem and $\theta = -\frac{\pi}{2}$. Making use of Euler's theorem to simplify $e^{\pm\frac{j\pi}{2}} = \cos\left(\frac{\pi}{2}\right) \pm j\sin\left(\frac{\pi}{2}\right) = \pm j$ and the fact that $-j = \frac{1}{j}$

$$S_{\text{mod}}(f) = \frac{1}{2}\left[e^{j\theta}S(f - f_c) + e^{-j\theta}S(f + f_c)\right] \tag{1.35}$$

$$= \frac{1}{2}\left[e^{\frac{-j\pi}{2}}\frac{T}{\left[1 + j2\pi\left(f - \frac{1}{T}\right)T\right]} + e^{\frac{j\pi}{2}}\frac{T}{\left[1 + j2\pi\left(f + \frac{1}{T}\right)T\right]}\right] \tag{1.36}$$

$$= \frac{T}{2}\left[-j\frac{1}{\left[1 + j2\pi\left(f - \frac{1}{T}\right)T\right]} + j\frac{1}{\left[1 + j2\pi\left(f + \frac{1}{T}\right)T\right]}\right] \tag{1.37}$$

$$= \frac{T}{2j}\left[\frac{1}{\left[1 + j2\pi T\left(f - \frac{1}{T}\right)\right]} - \frac{1}{\left[1 + j2\pi T\left(f + \frac{1}{T}\right)\right]}\right] \tag{1.38}$$

● ⭐ SIMULATION P11: For the rest of the solution. Experiment with the low-pass filter and observe the corresponding filtered signal.

12. The triangular pulse $s(t) = \begin{cases} A^2\left(1 - \frac{|t|}{T_b}\right) & \text{for} & |t| \le T_b \\ 0 & \text{otherwise} \end{cases}$

and its differentials $\frac{ds(t)}{dt}$ and $\frac{d^2 s(t)}{d^2 t}$ are shown below for $T_b = 0.5$ sec and $A = 1$ volt.

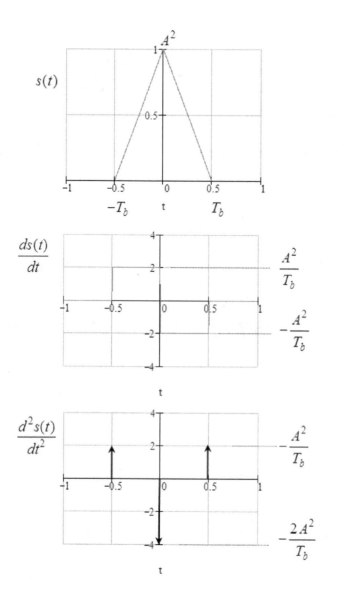

Note that $\frac{d^2 s(t)}{d^2 t}$ is an impulse at $\pm T_b$ and at $t = 0$ so that

$$\frac{d^2 s(t)}{dt^2} = \frac{A^2}{T_b} \delta\,(t - T_b) - \frac{2A^2}{T_b} \delta\,(t) + \frac{A^2}{T_b} \delta\,(t + T_b) \qquad (1.39)$$

$$= \frac{A^2}{T_b} \left[\delta\,(t - T_b) - 2\delta\,(t) + \delta\,(t + T_b) \right] \qquad (1.40)$$

From the time differentiation property

$$\frac{d^2 s(t)}{dt^2} \longleftrightarrow (j2\pi f)^2 \, S(f) = -(2\pi f)^2 \, S(f) \tag{1.41}$$

Bearing in mind that $\delta(t) \longleftrightarrow 1$ and making use of the time-shifting property, we have $\delta(t - T_b) \longleftrightarrow e^{-j2\pi f T_b}$. Thus

$$-(2\pi f)^2 \, S(f) = \frac{A^2}{T_b} \left[e^{-j2\pi f T_b} - 2 + e^{j2\pi f T_b} \right] \tag{1.42}$$

Now from Euler's theorem $\cos(2\pi f T_b) = \frac{1}{2}\left(e^{j2\pi f T_b} + e^{-j2\pi f T_b}\right)$, hence

$$-(2\pi f)^2 \, S(f) = \frac{2A^2}{T_b} \left[\cos(2\pi f T_b) - 1\right] = \frac{-2A^2}{T_b}\left[1 - \cos(2\pi f T_b)\right] \tag{1.43}$$

Making use of the trigonometric identity $\sin^2\theta = \frac{1}{2}(1 - \cos 2\theta)$ with $2\theta = 2\pi f T_b$

$$(2\pi f)^2 \, S(f) = \frac{2A^2}{T_b} \left[2\sin^2(\pi f T_b)\right] \tag{1.44}$$

$$S(f) = \frac{4A^2}{T_b} \left[\frac{\sin^2(\pi f T_b)}{(2\pi f)^2}\right] = A^2 T_b \left[\frac{\sin^2(\pi f T_b)}{(\pi f T_b)^2}\right] \tag{1.45}$$

but $\operatorname{sinc}(f T_b) = \frac{\sin(\pi f T_b)}{(\pi f T_b)}$, and therefore

$$S(f) = A^2 T_b \operatorname{sinc}^2(f T_b) \tag{1.46}$$

- $\leq$ ⭐ SIMULATION **P12**: The FT is determined and shown to be the same as $S(f) = A^2 T_b \operatorname{sinc}^2(f T_b)$. This simulation also contains the solution to part(b).

13. The FT of $s(t)$ is given by

$$S(f) = \int_{-\infty}^{\infty} s(t)e^{-j2\pi ft}dt. \tag{1.47}$$

Let $S_{at}(f)$ represent the FT of $s(at)$ i.e. $S_{at}(f) = \int_{-\infty}^{\infty} s(at)e^{-j2\pi ft}dt$. Let $\tau = at$ and assume $a > 0$. Then $\frac{d\tau}{dt} = a$ i.e. $dt = \left(\frac{1}{a}\right)d\tau$. Changing the integration to over $d\tau$, we have

$$S_{at}(f) = \frac{1}{a} \int_{-\infty}^{\infty} s(\tau)e^{-j2\pi\left(\frac{f}{a}\right)\tau}d\tau = \frac{1}{a}S\left(\frac{f}{a}\right) \tag{1.48}$$

and for $a < 0$,

$$S_{at}(f) = \frac{-1}{a} \int_{-\infty}^{\infty} s(\tau)e^{-j2\pi\left(\frac{f}{a}\right)\tau}d\tau = \frac{1}{|a|}S\left(\frac{f}{a}\right) \tag{1.49}$$

- ⭐ SIMULATION **P13:** To experiment with various scaling factors.

14. ⭐ SIMULATION **P14:** For the solution. Experiment with the frequency limits (fmin, fmax) of the filter and observe how the shape of the reconstructed signal is improved by allowing additional side-lobes to filter through.

Bandwidth of the pulse is infinite because the pulse in this example is rectangular. In practice, the edges of the pulse are "rounded-off" to minimize the spectral width of the pulse. Allowing the main lobe together with the first side-lobe is more than enough for a communication system because the objective is not to recover the pulse, but to establish the binary digits represented by the pulse.

15. Signal energy $= \displaystyle\int_{-\infty}^{\infty} |s(t)|^2 \, dt = \int_{0}^{0.5} dt = [t]_0^{0.5} = 0.5$ Joules.

• ⭐ SIMULATION **P15**: For the rest of the solution.   Experiment with the time duration of the pulse and notice the effect on its auto-correlation function.

16. ⭐ SIMULATION **P16**: For the full solution. Note that the auto-correlation function for this signal was derived in the text.

17. At time $t = 0$, capacitor is uncharged. Therefore, the charge on the plates $Q = 0$. Let V_R represent the voltage across the resistor at time t. Using Ohms law,

$$V_{in}(t) = V_R(t) + V_C(t) \tag{1.50}$$

where $V_R(t) = I(t)R$ and $V_C(t) = \frac{q(t)}{C}$, where $q(t)$ is the charge across the capacitor at time t. Thus,

$$V_{in}(t) = I(t)R + \frac{q(t)}{C} \tag{1.51}$$

Given that the current through the circuit $I(t) = \dfrac{dq(t)}{dt}$, we get

$$V_{in}(t) = R\frac{dq(t)}{dt} + \frac{q(t)}{C} \tag{1.52}$$

Since the input $V_{in}(t) = V_o$,

$$V_o = R\frac{dq(t)}{dt} + \frac{q(t)}{C} \tag{1.53}$$

Thus

$$CV_o - q = RC\frac{dq}{dt} \tag{1.54}$$

or equivalently,

$$\frac{1}{RC}\int_0^t dt = \int_0^Q \left(\frac{1}{CV_o - q}\right)dq \tag{1.55}$$

where Q is the charge on the capacitor after a time t. Let $u = CV_o - q$. Therefore, $\frac{du}{dq} = -1$. Changing the integration limits, for $q = 0, \ u = CV_o$ and for $q = Q, u = CV_o - Q$. Thus the new integral is

$$\frac{1}{RC}[t]_0^t = -\int_{CV_o}^{CV_o-Q} \left(\frac{1}{u}\right)du \tag{1.56}$$

$$-\frac{t}{RC} = [\log_e u]_{CV_o}^{CV_o-Q} = \log_e\left(\frac{CV_o - Q}{CV_o}\right) \tag{1.57}$$

$$e^{-\frac{t}{RC}} = \frac{CV_o - Q}{CV_o} \tag{1.58}$$

$$CV_o e^{-\frac{t}{RC}} = CV_o - Q \tag{1.59}$$

Thus the charge on the capacitor after time t is

$$Q(t) = CV_o - CV_o e^{-\frac{t}{RC}} = CV_o\left(1 - e^{-\frac{t}{RC}}\right) \tag{1.60}$$

The voltage across the capacitor after a time t is given by

$$V_{out}(t) = \frac{Q(t)}{C} = V_o\left(1 - e^{-\frac{t}{RC}}\right) \tag{1.61}$$

As expected for $t = \infty$, $V_{out(t)} = V_o$ i.e. the capacitor is charged fully.

18. (a)

$$S(f) = \int_{-\infty}^{\infty} s(t)e^{-j2\pi ft} dt \tag{1.62}$$

$$= \int_{-\infty}^{\infty} u(t)e^{-at}e^{-j2\pi ft} dt = \int_{0}^{\infty} e^{-at}e^{-j2\pi ft} dt \tag{1.63}$$

Notice the change of integration limit because $u(t) = \begin{cases} 1 & t > 0 \\ 0 & t < 0 \end{cases}$. Hence

$$S(f) = \int_{0}^{\infty} e^{-(a+j2\pi f)t} dt \tag{1.64}$$

$$= \left[-\frac{e^{-(a+j2\pi f)t}}{(a+j2\pi f)} \right]_{0}^{\infty} = \frac{1}{(a+j2\pi f)} \tag{1.65}$$

(b) The impulse response $h(t)$ of the RC filter is given by the inverse FT of

$$H(f) = \frac{1}{1 + j2\pi fRC} \tag{1.66}$$

$$= \frac{1}{RC\left(\dfrac{1}{RC} + j2\pi f\right)} = \frac{1}{RC}\left(\dfrac{1}{a + j2\pi f}\right) \tag{1.67}$$

where $a = \dfrac{1}{RC}$. Thus, using the result of part (a)

$$h(t) = \frac{1}{RC}\left(u(t)e^{-\frac{t}{RC}} \right) \tag{1.68}$$

Or equivalently, since $u(t) = \begin{cases} 1 & t > 0 \\ 0 & t < 0 \end{cases}$,

$$h(t) = \left(\frac{1}{RC} e^{-\frac{t}{RC}} \right) \; for \, t > 0 \qquad (1.69)$$

19. ⭐ SIMULATION **P19:** For the full solution.

20. For an RC filter,

$$H(f) = \frac{1}{1 + j2\pi f RC} = \frac{1}{1 + j2\pi f \left(\frac{1}{2\pi f c} \right)} \qquad (1.70)$$

since the filter cut-off frequency $fc = \left(\frac{1}{2\pi RC} \right)$. The input signal is periodic and so we need to determine its exponential Fourier Series.

• ⭐ SIMULATION **P20:** For the full solution. Experiment with different periodic input signals. Notice the curve at the end that shows how the output signal power depends on f_c. As expected, a larger f_c allows more power through the filter and saturates to the input power value.

21. ⭐ SIMULATION **P21:** For the full solution.

22. (a) The power in the signal $s(t) = A \sin \left(\frac{2\pi t}{T} \right)$ is given by

$$P = \frac{1}{T} \int_{-T/2}^{T/2} |s(t)|^2 \, dt \qquad (1.71)$$

$$= \frac{A^2}{T} \int_{-T/2}^{T/2} \sin^2 \left(\frac{2\pi t}{T} \right) dt = \frac{A^2}{2}$$

For $A = 3$ volts, $P = 4.5$ Watts.

(b) For the autocorrelation function $R(\tau) = \frac{A^2}{2}\cos\left(\frac{2\pi\tau}{T}\right)$, the PSD is given by the FT of $R(\tau)$, which in this case is simply two spikes on a double-sided spectrum at a frequency of $\frac{1}{T}$ Hz and amplitude $\left(\frac{\frac{A^2}{2}}{2}\right) = \frac{A^2}{4}$. Refer to the example in the text for full details.

(c) Total area under a PSD yields the signal power which in this case is simply $\frac{A^2}{4} + \frac{A^2}{4} = \frac{A^2}{2} = 4.5$ W as expected.

23. (a) Simply

$$s(t) = 5 + 6\cos\left(2\pi(10)t + \frac{150\pi}{180}\right) + 3\cos\left(2\pi(35)t - \frac{90\pi}{180}\right) \qquad (1.72)$$

(b) From the figure, $N = 32$. Assuming the sampling frequency is twice the maximum frequency i.e. $f_s = 2f_{max}$, its evident from the graph that $f_{max} = 40$ Hz and $f_s = 80$ Hz to yield the frequency resolution of $\frac{f_s}{N} = \frac{80}{32} = 2.5$ Hz. Alternatively, from the graph, notice that the frequency resolution is 2.5 Hz. Thus $f_s = 2.5(N) = 2.5(32) = 80$ Hz.

(c) Highest frequency is $f_{max} = \frac{f_2}{2} = 40$ Hz.

(d) There wold be a smearing effect near the spectral components at frequencies of $+10$ and $+35$ Hz. Similarly also on the negative half of the double-sided spectrum.

24. (a) The double-sided spectrum sketch is shown below, which is a periodic signal!

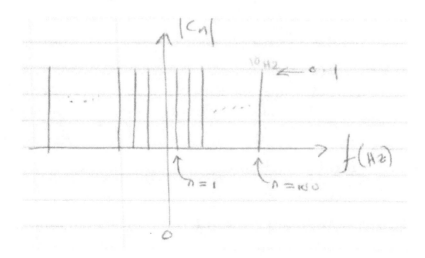

Recall from Parsevals theorem for a Fourier Series, that the power of a spectral component on a double-side spectrum is given by $|c_n|^2$. Thus power of this noise signal

$$P = |c_0|^2 + \sum_{n=1}^{\infty} 2\,|c_n|^2 \tag{1.73}$$

where the d.c component and all the components on the positive-half of the spectrum are used. But $c_0 = 0$ and therefore

$$P = \sum_{n=1}^{100} 2\,|0.1|^2 = 100(0.02) = 2W \tag{1.74}$$

(b) Using FFT, set the frequency resolution $\frac{f_s}{N}$ to match the frequency separation of spectral components, which is equal to $\frac{1}{10}$ Hz. Therefore $N = 10f_s$ for $f_s = 30$ Hz, $N = 300$.

(c) Sketch of the spectrum is shown below.

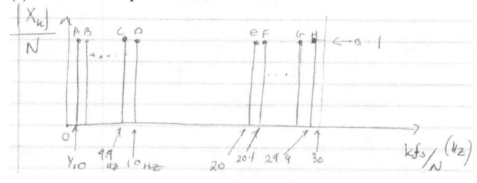

The maximum frequency is 0.1 (100) = 10 Hz. The first spectral component A corresponds to 0.1 Hz. Given that $f = \frac{kf_s}{N}$, $k = \frac{fN}{f_s} = \frac{0.1(300)}{30} = 1$. The second component B corresponds to 0.2 Hz with $k = 2$. The other components are C(9.9 Hz), D(10 Hz), E(20 Hz), F(20.1), G(29.9 Hz) and H(30 Hz). The value of k corresponding to each component is easily calculated using $k = f(10)$.

25. (a) The ensemble average

$$\overline{x_1} = \int_{-\infty}^{\infty} x_1 f_X(x_1) dx_1 = \int_{-\infty}^{\infty} A \cos{(2\pi f_o t_1 + \theta)} f_\theta(\theta) d\theta \qquad (1.75)$$

$$= \frac{2A}{\pi} \int_{-\frac{\pi}{4}}^{\frac{\pi}{4}} \cos{(2\pi f_o t_1 + \theta)} \, d\theta \qquad (1.76)$$

$$= \frac{2A}{\pi} \left[\sin(2\pi f_o t_1 + \frac{\pi}{4}) - \sin(2\pi f_o t_1 - \frac{\pi}{4}) \right]$$

$$= \frac{2A}{\pi} \left[\begin{array}{c} \sin(2\pi f_o t_1) \cos(\frac{\pi}{4}) + \cos(2\pi f_o t_1) \sin(\frac{\pi}{4}) \\ - \left\{ \sin(2\pi f_o t_1) \cos(\frac{\pi}{4}) - \cos(2\pi f_o t_1) \sin(\frac{\pi}{4}) \right\} \end{array} \right] \qquad (1.77)$$

and given that $\sin(\frac{\pi}{4}) = \cos(\frac{\pi}{4}) = \frac{1}{\sqrt{2}}$

$$\overline{x_1} = \frac{2\sqrt{2}A}{\pi} \cos(2\pi f_o t_1) \qquad (1.78)$$

Similarly,

$$\overline{x_1^2} = \int_{-\frac{\pi}{4}}^{\frac{\pi}{4}} [A \cos{(2\pi f_o t_1 + \theta)}]^2 \frac{2}{\pi} d\theta \qquad (1.79)$$

and making use of the identity $\cos^2 \theta = \frac{1}{2} (1 + \cos 2\theta)$

$$\overline{x_1^2} = \frac{A^2}{\pi} \int_{-\frac{\pi}{4}}^{\frac{\pi}{4}} d\theta + \frac{A^2}{\pi} \int_{-\frac{\pi}{4}}^{\frac{\pi}{4}} \cos\left(4\pi f_o t_1 + 2\theta\right) d\theta \tag{1.80}$$

$$= \frac{A^2}{\pi}\left(\frac{\pi}{2}\right) + \frac{A^2}{2\pi}\left[\sin\left(4\pi f_o t_1 + \frac{\pi}{2}\right) - \sin\left(4\pi f_o t_1 - \frac{\pi}{2}\right)\right] \tag{1.81}$$

but $\sin(\theta \pm \frac{\pi}{2}) = \pm \cos\theta$

$$\overline{x_1^2} = \frac{A^2}{2} + \frac{A^2}{\pi}\cos(4\pi f_o t_1) \tag{1.82}$$

Finally, the time average

$$\langle x^2(t) \rangle = \langle A^2 \cos^2\left(2\pi f_o t_1 + \theta\right) \rangle \tag{1.83}$$

$$= \frac{A^2}{2}\langle 1 + \cos\left(4\pi f_o t_1 + 2\theta\right) \rangle \tag{1.84}$$

$$= \frac{A^2}{2} + \frac{A^2}{2}\langle \cos\left(4\pi f_o t_1 + 2\theta\right) \rangle$$

By inspection, the time average value of the cos(.) function is zero. Thus $\overline{x_1^2} = \frac{A^2}{2}$. Since $\overline{x_1}$ is a function of time, the random process $x(t)$ is not stationary. Similarly $\overline{x_1^2}$ is a function of time and thus not a stationary random process. Furthermore, since $\overline{x_1^2} \neq \langle x^2(t) \rangle$, it is not ergodic.

(b) The PSD is given by the Fourier transform of

$$R_X(t_1, t_2) = \overline{x_1 x_2} = \overline{A \cos\left(2\pi f_o t_1 + \theta\right) A \cos\left(2\pi f_o t_2 + \theta\right)} \tag{1.85}$$

was shown to be given by

$$\frac{A^2}{2}\cos\left(2\pi f\tau\right) \tag{1.86}$$

Thus

$$PSD_X(f) = FT\left[\frac{A^2}{2}\cos{(2\pi f\tau)}\right] = \frac{A^2}{4}\left[\delta(f - f_o) + \delta(f + f_o)\right] \quad (1.87)$$

26. (a) We need to determine $\overline{r_1}$ at time t_1 to establish whether or not its a WSS random process. Thus

$$\overline{r_1} = \overline{A\cos{(2\pi f_o t_1 + \theta)} + n_1} \quad (1.88)$$

but

$$\overline{A\cos{(2\pi f_o t_1 + \theta)}} = A\cos{(2\pi f_o t_1 + \theta)} \quad (1.89)$$

because its not a random process and $\overline{n_1} = 0$

$$\therefore \overline{r_1} = A\cos{(2\pi f_o t_1 + \theta)} \quad (1.90)$$

is a function of time t_1 and therefore $r(t)$ is not a WSS process.

(b)

$$R_r(t_1, t_2) = \overline{(s_1 + n_1)(s_2 + n_2)} \quad (1.91)$$

$$= s_1 s_2 + s_1\overline{n_2} + s_2\overline{n_1} + \overline{n_1 n_2} \quad (1.92)$$

but $R_n(t_1, t_2) = \overline{n_1 n_2}$ and $\overline{n_1} = \overline{n_2} = 0$

$$\therefore \quad R_r(t_1, t_2) = s_1 s_2 + R_n(t_1, t_2) \quad (1.93)$$
$$= A^2\cos{(2\pi f_o t_1 + \theta)}\cos{(2\pi f_o t_2 + \theta)} + R_n(t_1, t_2)$$

27. Recall for a random process $z(t) = x(t) + y(t)$

• If $x(t)$ and $y(t)$ are **orthogonal**, $\overline{x_1 y_1} = 0$ and $P_Z = \overline{x_1^2} + \overline{y_1^2}$ is the sum of the power of $x(t)$ and $y(t)$.

- If $x(t)$ and $y(t)$ are **uncorrelated**, $\overline{x_1 y_1} = \overline{x_1}\,\overline{y_1}$ and $P_Z = \overline{x_1^2} + 2\,(\overline{x_1})\,(\overline{y_1}) + \overline{y_1^2}$.

In this case, $\overline{n_1^2} = \overline{n_1^2} = 4$ W and $\overline{n_1} = 1$ V and $\overline{n_2} = -1$ V. Thus, the power of $n(t)$ is $\overline{n_1^2} + \overline{n_2^2} = 4 + 4 = 8$ W if $n_1(t)$ and $n_2(t)$ are orthogonal and if they are uncorrelated, then $\overline{n_1^2} + 2\,(\overline{n_1})\,(\overline{n_2}) + \overline{n_2^2} = 4 + 2\,(1)\,(-1) + 4 = 6$ W.

28. The power spectral density (PSD) is given by the FT of the autocorrelation function. Thus

$$PSD(f) = FT\,[R_X(\tau)] = FT\left[A + Be^{-a|\tau|}\right] \tag{1.94}$$

$$= FT\,[A] + FT\left[Be^{-a|\tau|}\right] \tag{1.95}$$

Given the FT pairs $1 \leftrightarrow \delta(f)$ and $e^{-a|t|} \leftrightarrow \frac{2a}{a^2 + (2\pi f)^2}$

$$PSD(f) = A\delta(f) + B\frac{2a}{a^2 + (2\pi f)^2} \tag{1.96}$$

29. A plot of the power spectral density

$$PSD(f) = \begin{cases} \left(1 - \frac{|f|}{B}\right) & \text{for} \qquad |f| \le B \\ 0 & \text{otherwise} \end{cases} \tag{1.97}$$

is shown below for $B = 5$ Hz.

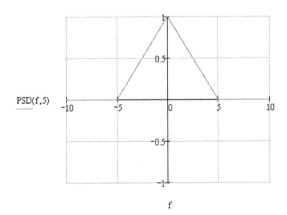

$$PSD(f,B) := \begin{vmatrix} 1 - \frac{|f|}{B} & \text{if } |f| \le B \\ 0 & \text{otherwise} \end{vmatrix}$$

(a) The average power of this random process is simply the area under the $PSD(f)$ curve, which in this case is given by
$P = \frac{1}{2}B + \frac{1}{2}B = B$ watts.

(b) Recall from question 12 that the FT of

$$s(t) = \begin{cases} A^2 \left(1 - \frac{|t|}{T_b}\right) & \text{for} \qquad |t| \le T_b \\ 0 & \text{otherwise} \end{cases} \tag{1.98}$$

is given by

$$S(f) = A^2 T_b \left[\frac{\sin^2(\pi f T_b)}{(\pi f T_b)^2}\right] \tag{1.99}$$

By comparison, and making use of the FT duality property, the autocorrelation function

$$R_X(\tau) = B \left[\frac{\sin^2(\pi \tau B)}{(\pi \tau B)^2}\right] \tag{1.100}$$

30. (a) The PSD of the sinusoid $A \sin(2\pi f_o t)$ from is given by

$$PSD(f) = \frac{A^2}{4} [\delta(f - f_o) + \delta(f + f_o)] \tag{1.101}$$

(b) The output PSD for the signal component is given by

$$PSD_{out}(f) = PSD_{in}(f)\,|H(f)|^2 \tag{1.102}$$

$$= \frac{A^2}{4}\,[\delta(f - f_o) + \delta(f + f_o)]\,|H(f)|^2$$

where $|H(f)|^2 = \begin{cases} \left(1 - 2\frac{|f|}{B} + \left(\frac{|f|}{B}\right)^2\right) & \text{for} \qquad |f| \le B \\ 0 & \text{otherwise} \end{cases}$ is shown

below for $B = 4$ Hz

$H(f,B) := \begin{vmatrix} 1 - \dfrac{|f|}{B} & \text{if } |f| \le B \\ 0 & \text{otherwise} \end{vmatrix}$

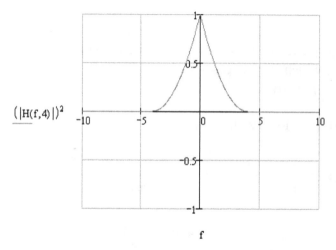

$(\,|H(f,4)\,|)^2$

Notice that

$$|H(f)|^2 = \begin{cases} \left(1 - 2\frac{f}{B} + \left(\frac{f}{B}\right)^2\right) & \text{for} \quad 0 \le f \le B \\ \left(1 + 2\frac{f}{B} + \left(\frac{f}{B}\right)^2\right) & \text{for} \quad -B \le f \le 0 \end{cases} \tag{1.103}$$

(c) Given that B is greater than the frequency f_o of the sinusoid, the output signal power

$$P_{signal} = \int_{-\infty}^{\infty} PSD_{out}(f)\,df \tag{1.104}$$

$$= \frac{A^2}{4} \int_0^B \delta(f - f_o) \, |H(f)|^2 \, df + \frac{A^2}{4} \int_{-B}^0 \delta(f + f_o) \, |H(f)|^2 \, df \qquad (1.105)$$

But $\int_0^B \delta(f - f_o)df = 1$ only when $f = f_o$ and $\int_{-B}^0 \delta(f + f_o)df = 1$ only when $f = -f_o$, otherwise its zero. Thus its sifts out the values

$$P_{signal} = \frac{A^2}{4} |H(f_0)|^2 + \frac{A^2}{4} |H(-f_o)|^2 \qquad (1.106)$$

$$= \frac{A^2}{4} \left[\left(1 - 2\frac{f_o}{B} + \left(\frac{f_0}{B}\right)^2\right) + \left(1 + 2\frac{(-f_0)}{B} + \left(\frac{-f_0}{B}\right)^2\right) \right] \qquad (1.107)$$

$$= \frac{A^2}{4} \left[1 - 2\frac{f_0}{B} + \left(\frac{f_0}{B}\right)^2 + 1 - 2\frac{f_o}{B} + \left(\frac{f_0}{B}\right)^2 \right] \qquad (1.108)$$

$$= \frac{2A^2}{4} \left[1 - 2\frac{f_0}{B} + \left(\frac{f_0}{B}\right)^2 \right] = \frac{A^2}{2} \left[1 - 2\frac{f_0}{B} + \left(\frac{f_0}{B}\right)^2 \right] \qquad (1.109)$$

For example, if $A = 1$, $f_o = 1$ Hz and $B = 4$ Hz, then

$$P_{signal} = \frac{1}{2} \left[1 - 2\frac{1}{4} + \left(\frac{1}{4}\right)^2 \right] = \frac{9}{32} \text{Watts} \qquad (1.110)$$

(d) The output PSD for the noise component is given by

$$PSD_{out}(f) = PSD_{in}(f) \, |H(f)|^2 = \frac{N_o}{2} |H(f)|^2 \qquad (1.111)$$

where

$$(1.112)$$

$$|H(f)|^2 \;=\; \begin{cases} \left(1 - 2\frac{f}{B} + \left(\frac{f}{B}\right)^2\right) & \text{for} \quad 0 \le f \le B \quad \text{positive f} \\[2mm] \left(1 + 2\frac{f}{B} + \left(\frac{f}{B}\right)^2\right) & \text{for} \quad -B \le f \le 0 \quad \text{negative f} \end{cases}$$

The average noise power output is given by

$$P_{noise} = \int_{-\infty}^{\infty} PSD_{out}(f)\,df \qquad (1.113)$$

$$= \frac{N_o}{2} \int_{0}^{B} \left(1 - 2\frac{f}{B} + \left(\frac{f}{B}\right)^2\right) df + \frac{N_o}{2} \int_{-B}^{0} \left(1 + 2\frac{f}{B} + \left(\frac{f}{B}\right)^2\right) df \quad (1.114)$$

$$= \frac{N_o}{2}\left[f - \frac{f^2}{B} + \frac{f^3}{3B^2}\right]_0^{B} + \frac{N_o}{2}\left[f + \frac{f^2}{B} + \frac{f^3}{3B^2}\right]_{-B}^{0} \qquad (1.115)$$

$$= \frac{N_o}{2}\left[\left(B - B + \frac{B}{3}\right) - 0\right] + \frac{N_o}{2}\left[0 - \left(-B + B - \frac{B}{3}\right)\right] \qquad (1.116)$$

$$= \frac{N_o}{2}\left(\frac{B}{3} + \frac{B}{3}\right) = \frac{N_o B}{3} \qquad (1.117)$$

For example if $N_o = 10^{-12}$ W/Hz and $B = 4$ Hz, then

$$P_{noise} = \frac{10^{-12}}{12}\text{W} \qquad (1.118)$$

Chapter 2

Analog Communications

2.1 Solutions

1. (a) Let

$$y(t) = s_{DSB}(t) \cos\left(2\pi f_c t + \phi\right) \tag{2.1}$$

(i) We have

$$y(t) = m(t) \cos(2\pi f_c t) \cos\left(2\pi f_c t + \phi\right) \tag{2.2}$$

$$= m(t) \cos(2\pi f_c t) \left[\cos(2\pi f_c t) \cos\phi - \sin(2\pi f_c t) \sin\phi\right]$$

$$= m(t) \cos\phi \cos^2(2\pi f_c t) - m(t) \sin\phi \cos(2\pi f_c t) \sin(2\pi f_c t) \tag{2.3}$$

$$= \frac{m(t) \cos\phi}{2} \left[\cos(4\pi f_c t) + 1\right] - \frac{m(t) \sin\phi}{2} \sin(4\pi f_c t)$$

$$= \frac{m(t) \cos\phi}{2} + \frac{m(t) \cos\phi}{2} \cos(4\pi f_c t) - \frac{m(t) \sin\phi}{2} \sin(4\pi f_c t)$$

$$= \frac{m(t) \cos\phi}{2} + \frac{m(t)}{2} \cos(4\pi f_c t + \phi)$$

Using a low-pass filter to eliminate the high frequency term $\frac{m(t)}{2} \cos(4\pi f_c t + \phi)$

27

$$y(t) = \frac{m(t)\cos\phi}{2}. \ (ii) For \phi = \pm\frac{\pi}{2}, x(t) = 0 \tag{2.4}$$

(b) Simply

$$
\begin{aligned}
y(t) &= m(t)\cos(2\pi f_c t)\cos[2\pi(f_c + f)t] &\tag{2.5}\\
&= m(t)\cos(2\pi f_c t)\cos(2\pi f_c t + 2\pi f t) &\tag{2.6}
\end{aligned}
$$

Comparing this expression with that in part(a), we may let $\phi = 2\pi f t$. Following the same analysis as in part(a),

$$y(t) = \frac{m(t)\cos 2\pi f t}{2} + \frac{m(t)}{2}\cos(4\pi f_c t + 2\pi f t) \tag{2.7}$$

After the low-pass filter, we are left with the term

$$y(t) = \frac{m(t)\cos 2\pi f t}{2} \tag{2.8}$$

Ideally, $f = 0$ and $y(t) = \frac{m(t)}{2}$.

(c)   The effect of a phase error ϕ is simply is reduction of $\frac{m(t)}{2}$ by a factor of $\cos\phi$. If ϕ remains constant, then it effect is easily overcome by amplification.   However, a frequency error f in the local oscillator results in the message signal $m(t)$ being multiplied by $\cos 2\pi f t$ which is difficult to overcome i.e. its a nightmare!

2. The power of a periodic signal $s_{DSB}(t) = m(t)A_c\cos(2\pi f_c t)$ is given by

$$P_{DSB} = \frac{1}{T}\int_{-\frac{T}{2}}^{\frac{T}{2}} s_{DSB}^2(t)dt \tag{2.9}$$

$$= \frac{A_c^2}{T} \int_{-\frac{T}{2}}^{\frac{T}{2}} m^2(t) \cos^2(2\pi f_c t)\, dt = \frac{A_c^2}{2T} \int_{-\frac{T}{2}}^{\frac{T}{2}} m^2(t)\,[1 + \cos(4\pi f_c t)]\, dt \qquad (2.10)$$

$$= \frac{A_c^2}{2T} \int_{-\frac{T}{2}}^{\frac{T}{2}} m^2(t)\,dt + \frac{A_c^2}{2T} \int_{-\frac{T}{2}}^{\frac{T}{2}} m^2(t) \cos(4\pi f_c t)\, dt \qquad (2.11)$$

Given that the average value of the function $\cos(4\pi f_c t)$ is zero, we find

$$P_{DSB} = \frac{A_c^2}{2T} \int_{-\frac{T}{2}}^{\frac{T}{2}} m^2(t)\,dt = \frac{A_c^2 P_m}{2} \qquad (2.12)$$

3. ★ SIMULATION **P3:** For the full solution.

- Try a different expression for $m(t)$.

4. ★ SIMULATION **P4:** For the full solution.

5. We have

$$s_{AM}(t) = m(t) \cos(2\pi f_c t) + A_c \cos(2\pi f_c t) \qquad (2.13)$$

$$m(t) = 2 \sin(2\pi t) \qquad (2.14)$$

with $\mu = \frac{|m(t)_{\min}|}{A_c} = \frac{2}{A_c} = 0.2$ and $A_c = \frac{2}{0.2} = 10$ volts.

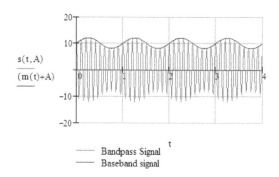

- ⭐ SIMULATION **P5:** For further details.   Observe the modulated carrier for various values of μ, in particular try 0.5 and 1.   Try a different expression for $m(t)$.

6. (a) The power of a periodic signal $s_{AM}(t) = [m(t) + A_c] \cos(2\pi f_c t)$ is given by

$$P_{AM} = \frac{1}{T} \int_{-\frac{T}{2}}^{\frac{T}{2}} s_{AM}^2(t) dt \tag{2.15}$$

$$= \frac{1}{T} \int_{-\frac{T}{2}}^{\frac{T}{2}} [m(t) + A_c]^2 \cos^2(2\pi f_c t) \, dt \tag{2.16}$$

$$= \frac{1}{2T} \int_{-\frac{T}{2}}^{\frac{T}{2}} (m(t) + A_c)(m(t) + A_c)[1 + \cos(4\pi f_c t)] \, dt \tag{2.17}$$

$$= \frac{1}{2T} \int_{-\frac{T}{2}}^{\frac{T}{2}} (m^2(t) + 2A_c m(t) + A_c^2)[1 + \cos(4\pi f_c t)] \, dt \tag{2.18}$$

$$= \frac{1}{2T} \int_{-\frac{T}{2}}^{\frac{T}{2}} \left[\begin{matrix} m^2(t) + m^2(t) \cos(4\pi f_c t) + 2Am(t) \\ +2Am(t) \cos(4\pi f_c t) + A^2 + A^2 \cos(4\pi f_c t) \end{matrix} \right] dt \qquad (2.19)$$

All the terms with $\cos(4\pi f_c t)$ will go to zero over the time period T. Hence

$$P_{AM} = \frac{1}{2T} \int_{-\frac{T}{2}}^{\frac{T}{2}} \left[m^2(t) + 2A_c m(t) + A_c^2 \right] dt \qquad (2.20)$$

$$= \frac{1}{2T} \int_{-\frac{T}{2}}^{\frac{T}{2}} m^2(t) dt + \frac{2A}{2T} \int_{-\frac{T}{2}}^{\frac{T}{2}} m(t) dt + \frac{A_c^2}{2T} T \qquad (2.21)$$

Assuming the average value of $m(t)$ is zero, then

$$\frac{1}{T} \int_{-\frac{T}{2}}^{\frac{T}{2}} m(t) dt = 0 \qquad (2.22)$$

$$P_{AM} = \frac{P_m}{2} + \frac{A_c^2}{2} = \frac{(P_m + A_c^2)}{2} \qquad (2.23)$$

(b) ⭐ SIMULATION **P6:** For part(b).

7. (a) For the coherent demodulation of a conventional amplitude modulation signal given by $s_{AM}(t) = [m(t) + A_c] \cos(2\pi f_c t)$, we have

$$y(t) = s_{AM}(t) \cos(2\pi f_c t) = [m(t) + A_c] \cos^2(2\pi f_c t) \qquad (2.24)$$

$$= [m(t) + A_c] \frac{[1 + \cos(4\pi f_c t)]}{2} = \left(\frac{m(t)}{2} + \frac{A_c}{2}\right)[1 + \cos(4\pi f_c t)] \quad (2.25)$$

$$= \frac{m(t)}{2} + \frac{A_c}{2} + \left(\frac{m(t)}{2} + \frac{A_c}{2}\right)\cos(4\pi f_c t)$$

which when passed through a low-pass filter, leaves $y(t) = \frac{m(t)}{2} + \frac{A}{2}$ from which $m(t)$ is easily extracted by simply removing the d.c. term $\frac{A_c}{2}$ and amplifying the signal by a factor of 2.

(b) ⭐ SIMULATION **P7:** For part(b). The corresponding simulation results are shown below.

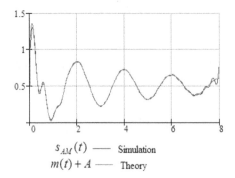

$s_{AM}(t)$ —— Simulation
$m(t) + A$ —— Theory

8. (a)

$$s_{AM}(t) = [m(t) + A_c]\cos(2\pi f_c t) = A_c \left[\frac{m(t)}{A_c} + 1\right]\cos(2\pi f_c t) \quad (2.26)$$

where

$$m(t) = A_m \cos(2\pi f_m t) \quad (2.27)$$

Given that $\mu = \frac{|m(t)_{\min}|}{A_c} = \frac{A_m}{A_c}$

$$
\begin{aligned}
s_{AM}(t) &= [A_m \cos(2\pi f_m t) + A_c]\cos(2\pi f_c t) & (2.28)\\
&= [\mu A_c \cos(2\pi f_m t) + A_c]\cos(2\pi f_c t) & (2.29)\\
&= [1 + \mu \cos(2\pi f_m t)]A_c \cos(2\pi f_c t) & (2.30)
\end{aligned}
$$

(b). ⭐ SIMULATION **P8:** For part(b).

9. ⭐ SIMULATION **P9:** For the full solution. Key results are shown below.

(a)

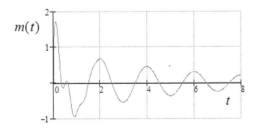

(b) 3.7 volts.

(c) Now $v_{cap}(t) = V_{\max}e^{-\frac{t}{t_c}}$. Using the first two terms of the Taylor series

$$v_{approx}(t) = v_{cap}(a) + (t - a)\frac{dv_{cap}(a)}{dt} \qquad (2.31)$$

where

$$\frac{dv_{cap}(t)}{dt} = -\frac{V_{\max}}{t_c}e^{-\frac{t}{t_c}} \qquad (2.32)$$

For $a = 0$, $\frac{dv_{cap}(0)}{dt} = -\frac{V_{\max}}{t_c}$. Thus

$$v_{approx}(t) = V_{\max} - t\frac{V_{\max}}{t_c} = V_{\max}\left(1 - \frac{t}{t_c}\right) \qquad (2.33)$$

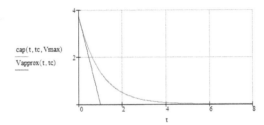

(d)

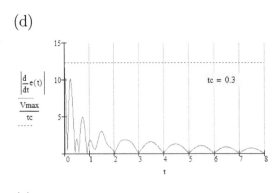

(e)

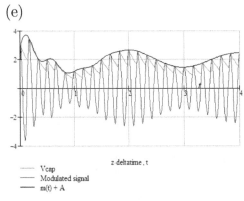

(f)

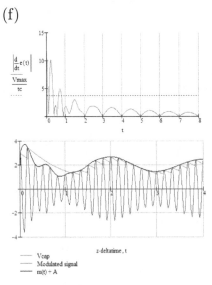

10. For $m(t) = A_m \cos(2\pi f_m t)$

$$m_p(t) = A_m \cos(2\pi f_m t - \frac{\pi}{2}) \qquad (2.34)$$

$$= A_m \cos 2\pi f_m t \cos \frac{\pi}{2} + A_m \sin 2\pi f_m t \sin \frac{\pi}{2} = A_m \sin 2\pi f_m t \qquad (2.35)$$

$$
\begin{aligned}
s_{USB}(t) &= A_m \cos(2\pi f_m t) \cos(2\pi f_c t) - A_m \sin 2\pi f_m t \sin(2\pi f_c t) \qquad (2.36)\\
&= A_m \cos[2\pi (f_m + f_c) t] \qquad (2.37)
\end{aligned}
$$

$$
\begin{aligned}
s_{LSB}(t) &= A_m \cos(2\pi f_m t) \cos(2\pi f_c t) + A_m \sin 2\pi f_m t \sin(2\pi f_c t) \qquad (2.38)\\
&= A_m \cos[2\pi (f_m - f_c) t]
\end{aligned}
$$

11. (a) For the coherent demodulation of a USB signal given by

$$s_{USB}(t) = m(t) \cos(2\pi f_c t) - m_p(t) \sin(2\pi f_c t) \qquad (2.39)$$

we have

$$y(t) = s_{USB}(t) \cos(2\pi f_c t) = m(t) \cos^2(2\pi f_c t) - m_p(t) \sin(2\pi f_c t) \cos(2\pi f_c t) \qquad (2.40)$$

$$
\begin{aligned}
&= \frac{m(t)}{2} [\cos(4\pi f_c t) + 1] - \frac{m_p(t)}{2} \sin(4\pi f_c t) \qquad (2.41)\\
&= \frac{m(t)}{2} + \frac{m(t)}{2} \cos(4\pi f_c t) - \frac{m_p(t)}{2} \sin(4\pi f_c t)
\end{aligned}
$$

which when passed through a low-pass filter, leaves $y(t) = \frac{m(t)}{2}$. Similarly for $s_{LSB}(t)$, we find that

$$y(t) = \frac{m(t)}{2} + \frac{m(t)}{2} \cos(4\pi f_c t) + \frac{m_p(t)}{2} \sin(4\pi f_c t) \qquad (2.42)$$

from which we are left with $\frac{m(t)}{2}$ after the low-pass filter.

- ⭐ SIMULATION **P11:** For part(b).

12. (a) From

$$s_{USB}(t) = m(t) \cos(2\pi f_c t) - m_p(t) \sin(2\pi f_c t) \qquad (2.43)$$

$$
\begin{aligned}
y_1(t) &= m(t) \cos\left(2\pi f_c t - \frac{\pi}{2}\right) - m_p(t) \sin\left(2\pi f_c t - \frac{\pi}{2}\right) \qquad (2.44) \\
&= m(t) \cos(2\pi f_c t) \cos\left(\frac{\pi}{2}\right) + m(t) \sin(2\pi f_c t) \sin\left(\frac{\pi}{2}\right) \\
&\quad -m_p(t)\left[\sin(2\pi f_c t) \cos\left(\frac{\pi}{2}\right) - \cos(2\pi f_c t) \sin\left(\frac{\pi}{2}\right)\right] \\
&= m(t) \sin(2\pi f_c t) - m_p(t)\left[-\cos(2\pi f_c t)\right] \\
&= m(t) \sin(2\pi f_c t) + m_p(t) \cos(2\pi f_c t)
\end{aligned}
$$

(b) Similarly

$$y_2(t) = s_{USB}(t) \cos(2\pi f_c t) + y_1(t) \sin(2\pi f_c t) \qquad (2.45)$$

$$
\begin{aligned}
&= m(t) \cos^2(2\pi f_c t) - m_p(t) \sin(2\pi f_c t) \cos(2\pi f_c t) \qquad (2.46) \\
&\quad +m(t) \sin^2(2\pi f_c t) + m_p(t) \cos(2\pi f_c t) \sin(2\pi f_c t) \\
&= m(t)\left[\cos^2(2\pi f_c t) + \sin^2(2\pi f_c t)\right] \\
&= m(t)
\end{aligned}
$$

13. ⭐ SIMULATION **P13:** For the solution. Note the excellent recovery of $m(t)$.

14. Now

$$s_{QAM}(t) = m_1(t) \cos(2\pi f_c t) + m_2(t) \sin(2\pi f_c t) \qquad (2.47)$$

(a) Simply

$$y(t) = s_{QAM}(t) \cos(2\pi f_c t) \tag{2.48}$$

$$
\begin{aligned}
&= m_1(t) \cos^2(2\pi f_c t) + m_2(t) \sin(2\pi f_c t) \cos(2\pi f_c t) &\text{(2.49)}\\
&= \frac{m_1(t)}{2}[\cos(4\pi f_c t) + 1] + \frac{m_2(t)}{2}\sin(4\pi f_c t)\\
&= \frac{m_1(t)}{2} + \frac{m_1(t)}{2}\cos(4\pi f_c t) + \frac{m_2(t)}{2}\sin(4\pi f_c t)
\end{aligned}
$$

which after the use of a low-pass filter

$$y(t) = \frac{m_1(t)}{2} \tag{2.50}$$

(b) Similarly

$$y(t) = s_{QAM}(t) \sin(2\pi f_c t) \tag{2.51}$$

$$
\begin{aligned}
&= m_1(t) \cos(2\pi f_c t) \sin(2\pi f_c t) + m_2(t) \sin^2(2\pi f_c t) &\text{(2.52)}\\
&= \frac{m_1(t)}{2}\sin(4\pi f_c t) + \frac{m_2(t)}{2}[1 - \cos(4\pi f_c t)]\\
&= \frac{m_2(t)}{2} + \frac{m_1(t)}{2}\sin(4\pi f_c t) - \frac{m_2(t)}{2}\cos(4\pi f_c t)
\end{aligned}
$$

which after the use of a low-pass filter

$$y(t) = \frac{m_2(t)}{2} \tag{2.53}$$

(c) ⭐ SIMULATION **P14:** For the solution. The demodulated signals $m_1(t)$ and $m_2(t)$ are shown below. Experiment with different message signals, f_c, etc. i.e. have some fun with the code!

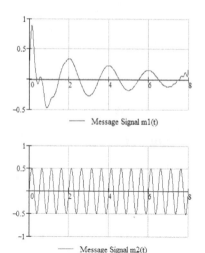

— Message Signal m1(t)

— Message Signal m2(t)

15. For $s(t) = m(t) \cos \left[2\pi \left(f_c - 2f_I \right) t \right]$, the spectrum of $y(t)$ will consist of side-bands centered on

$$f_L + (f_c - 2f_I) = f_L + (f_L + f_I) - 2f_I = 2f_L - f_I \tag{2.54}$$

$$f_L - (f_c - 2f_I) = f_L - f_c + 2f_I = f_L - (f_L + f_I) + 2f_I = f_I \tag{2.55}$$

i.e. the message signal carried by the image frequency $f_{image} = f_c - 2f_I$ will also pass through the band-pass filter centered on f_I if it is not filtered out prior to mixing.

16. Given $f_c = 900$ kHz and $f_I = 455$ kHz

Via up-conversion,

$$
\begin{aligned}
f_L &= f_c + f_I \\
f_L &= 1355 \text{ kHz}
\end{aligned}
\tag{2.56}
$$

Via down-conversion

$$f_L = f_c - f_I \qquad (2.57)$$
$$f_L = 900 - 455 = 445 \text{ kHz}$$

17. (a) For $f_{\min} \le f_c \le f_{\max}$, we find that for $f_L = f_c - f_I$, $f_{L\min} = f_{\min} - f_I$ and $f_{L\max} = f_{\max} - f_I$, so that

$$\left(\frac{f_{L\max}}{f_{L\min}} \right)_{Down} = \frac{f_{\max} - f_I}{f_{\min} - f_I} \qquad (2.58)$$

Similarly,

$$f_L = f_c + f_I, \quad \left(\frac{f_{L\max}}{f_{L\min}} \right)_{Up} = \frac{f_{\max} + f_I}{f_{\min} + f_I} \qquad (2.59)$$

For $f_{\min} = 550$ kHz, $f_{\max} = 1600$ kHz and $f_I = 455$ kHz

$$\left(\frac{f_{L\max}}{f_{L\min}} \right)_{Down} = 12.05 \qquad (2.60)$$
$$\left(\frac{f_{L\max}}{f_{L\min}} \right)_{Up} = 2.045$$

Clearly, up-conversion requires a smaller ratio for $\frac{f_{L\max}}{f_{L\min}}$.

(b) ⭐ SIMULATION **P17:** For the graph of $\frac{f_{L\max}}{f_{L\min}}$ versus f_I. Notice how $\frac{f_{L\max}}{f_{L\min}}$ is reduced as f_I is increased.

18. ⭐ SIMULATION **P18:** For the full solution.

19. Given

$$\theta(t) = (2\pi f_c t + \phi(t)) \tag{2.61}$$

(a) For $\phi(t) = \frac{\pi t^2}{2}$, the instantaneous frequency ν is given by

$$\nu = \frac{1}{2\pi}\frac{d\theta(t)}{dt} = f_c + \frac{t}{2} \tag{2.62}$$

(b) For $\phi(t) = \frac{1}{2}\sin(4\pi t)$,

$$\nu = \frac{1}{2\pi}\frac{d\theta(t)}{dt} = \frac{1}{2\pi}\left(2\pi f_c + 2\pi \cos(4\pi t)\right) = f_c + \cos(4\pi t) \tag{2.63}$$

i.e. frequency deviation is given by $\cos(4\pi t)$, from which its clear that the maximum frequency deviation is 1 Hz. From the expression $\phi(t) = \frac{1}{2}\sin(4\pi t)$, given that the maximum value of $\sin(.)$ is $+1$, the maximum phase deviation is simply $\frac{1}{2}$.

- ⭐ SIMULATION **P19:** For further details. Try different expressions for $\phi(t)$.

20. ⭐ SIMULATION **P20:** For the full solution.

21. Given

$$
\begin{aligned}
m(t) &= A_m \cos(2\pi f_m t) \tag{2.64}\\
s(t) &= A_c \cos(2\pi f_c t + \phi(t))
\end{aligned}
$$

$$\frac{d\phi(t)}{dt} = k_f m(t) \tag{2.65}$$

$$\theta(t) = 2\pi f_c t + \phi(t) \tag{2.66}$$

(a) Instantaneous frequency

$$
\begin{aligned}
\nu &= \frac{1}{2\pi}\frac{d\theta(t)}{dt} = \frac{1}{2\pi}\left[2\pi f_c + \frac{d\phi(t)}{dt}\right] = f_c + \frac{1}{2\pi}\frac{d\phi(t)}{dt} \tag{2.67}\\
&= f_c + \frac{k_f A_m}{2\pi}\cos\left(2\pi f_m t\right)
\end{aligned}
$$

(b) The frequency deviation is

$$\frac{k_f A_m}{2\pi}\cos\left(2\pi f_m t\right) \tag{2.68}$$

from which the maximum frequency deviation

$$\Delta f = \frac{k_f A_m}{2\pi} \tag{2.69}$$

(c) Hence deviation ratio

$$D = \frac{\Delta f}{f_m} = \frac{k_f A_m}{2\pi f_m} \tag{2.70}$$

(d) Bandwidth is simply given by

$$B_{FM} = 2\left(D+1\right)f_m = 2\left(\frac{k_f A_m}{2\pi f_m}+1\right)f_m \tag{2.71}$$

(e) Bandwidth

$$B_{FM} = 2\left(\frac{k_f A_m}{2\pi f_m}+1\right)f_m = 2\left(\frac{30}{4\pi}+1\right)2 = 13.5 \text{ Hz} \tag{2.72}$$

$$Power = \frac{A_c^2}{2} = 50 \text{ watts} \tag{2.73}$$

● ★ SIMULATION **P21**: For further details. Experiment with different expressions for $m(t)$.

22. Given

$$s(t) = A_c \cos\left(2\pi f_c t + \phi(t)\right) \tag{2.74}$$
$$m(t) = A_m \sin\left(2\pi f_m t\right)$$
$$\phi(t) = k_p m(t)$$

$$\frac{d\phi(t)}{dt} = k_p \frac{dm(t)}{dt} = k_p A_m 2\pi f_m \cos\left(2\pi f_m t\right) \tag{2.75}$$

$$\theta(t) = 2\pi f_c t + \phi(t) \tag{2.76}$$

(a) Instantaneous frequency

$$\begin{aligned}
\nu &= \frac{1}{2\pi}\frac{d\theta(t)}{dt} = \frac{1}{2\pi}\left[2\pi f_c + k_p A_m 2\pi f_m \cos\left(2\pi f_m t\right)\right] \\
&= f_c + k_p A_m f_m \cos\left(2\pi f_m t\right)
\end{aligned} \tag{2.77}$$

(b) The frequency deviation is

$$k_p A_m f_m \cos\left(2\pi f_m t\right) \tag{2.78}$$

from which the maximum frequency deviation

$$\Delta f = k_p A_m f_m \tag{2.79}$$

(c) Hence deviation ratio

$$D = \frac{\Delta f}{f_m} = \frac{k_p A_m f_m}{f_m} = k_p A_m \tag{2.80}$$

(d) Bandwidth

$$B_{PM} = 2\left(D+1\right) f_m = 2\left(k_p A_m + 1\right) f_m \tag{2.81}$$

(e) Bandwidth

$$B_{PM} = 2\left(k_p A_m + 1\right) f_m = 2\left(30+1\right)1 = 62 \text{ Hz} \tag{2.82}$$

$$Power = \frac{A_c^2}{2} = 50 \text{ watts} \tag{2.83}$$

● ★ SIMULATION **P22:** For further details. Experiment with different expressions for $m(t)$.

23. Given

$$\phi(t) = k_p m(t) \tag{2.84}$$

$$
\begin{aligned}
s(t) &= A_c \cos\left(2\pi f_c t + \phi(t)\right) &&(2.85)\\
&= A_c \cos\left[2\pi f_c t + k_p A_m \sin\left(2\pi f_m t\right)\right]\\
&= A_c \cos\left(2\pi f_c t\right) \cos\left(k_p A_m \sin\left(2\pi f_m t\right)\right)\\
&\quad - A_c \sin\left(2\pi f_c t\right) \sin\left(k_p A_m \sin\left(2\pi f_m t\right)\right)
\end{aligned}
$$

From the previous problem, we note that in this case, $D = k_p A_m$. Hence

$$
\begin{aligned}
s(t) &= A_c \cos\left(2\pi f_c t\right) \cos\left(D \sin\left(2\pi f_m t\right)\right) &&(2.86)\\
&\quad - A_c \sin\left(2\pi f_c t\right) \sin\left(D \sin\left(2\pi f_m t\right)\right)
\end{aligned}
$$

For $D \ll 1$,

$$\sin\left(D \sin\left(2\pi f_m t\right)\right) \approx D \sin\left(2\pi f_m t\right) \; and \cos\left(D \sin\left(2\pi f_m t\right)\right) \approx 1 \tag{2.87}$$

Thus

$$s(t) = A_c \cos\left(2\pi f_c t\right) - A_c \sin\left(2\pi f_c t\right) D \sin\left(2\pi f_m t\right) \tag{2.88}$$

Making use of the trigonometric identity

$$\sin A \sin B = \frac{1}{2}\cos(A - B) - \frac{1}{2}\cos(A + B) \tag{2.89}$$

$$s(t) = A_c \cos\left(2\pi f_c t\right) - \frac{A_c D}{2} \cos\left[2\pi(f_c - f_m)t\right] + \frac{A_c D}{2} \cos\left[2\pi(f_c + f_m)t\right]$$

$$(2.90)$$

i.e. spectrum contains three lines at f_c, $(f_c - f_m)$ and $(f_c + f_m)$ of height $\frac{A_c}{2}$, $\frac{A_c D}{4}$ and $\frac{A_c D}{4}$ on a double-sided spectrum.

● ⭐ SIMULATION **P23:** For the solution to part(b).   The spectra for $k_p = 0.1$ and 5 are shown below.   Modify the code to repeat this problem, but now for the FM signal in problem 21.

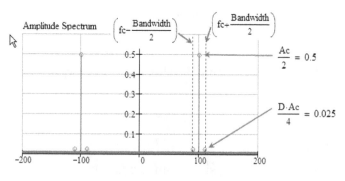

$$k_p = 0.1.$$

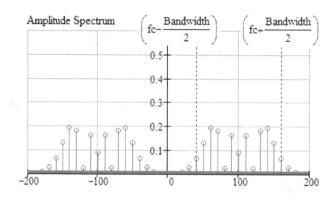

$$k_p = 5.$$

24. (a) Simply

$$s(t) = A_c \cos\left(2\pi f_c t + 100 \cos\left(2\pi t\right)\right) \tag{2.91}$$

from which

$$\theta(t) = 2\pi f_c t + 100 \cos\left(2\pi t\right) \tag{2.92}$$

The instantaneous frequency

$$\begin{aligned}\nu &= \frac{1}{2\pi}\frac{d\theta(t)}{dt} = \frac{1}{2\pi}\left[2\pi f_c - 2\pi 100 \sin\left(2\pi t\right)\right] \\ &= f_c - 100 \sin\left(2\pi t\right)\end{aligned} \tag{2.93}$$

Thus the maximum frequency deviation $\Delta f = 100$ Hz. Note that $s(t)$ could either be an FM or a PM signal. All we know is that the message signal is a tone signal with frequency $f_m = 1$ Hz. Thus the deviation ratio $D = \frac{\Delta f}{f_m} = 100$.

(b) The bandwidth

$$2\left(D + 1\right) f_m = 202 \text{ Hz} \tag{2.94}$$

Given that $D \gg 1$, $s(t)$ is a wideband angle-modulated signal for which the bandwidth is approximately given by $2\Delta f = 200$ Hz.

(c) ⭐ SIMULATION **P24:** For the solution.

25. ⭐ SIMULATION **P25:** For the full solution. Experiment with different expressions for $m(t)$.

26. Following the analysis in the text,

$$Z_1 = \frac{1}{j2\pi fC}, \, Z_2 = R \qquad\qquad (2.95)$$

$$\frac{V_{out}}{V_{in}} = \frac{Z_2}{(Z_1 + Z_2)} = \frac{R}{\frac{1}{j2\pi fC} + R} = \frac{j2\pi fRC}{1 + j2\pi fRC} \qquad (2.96)$$

Given that time constant $t_c = RC$, the frequency response $H(f)$ of the filter is given by

$$H(f) = \frac{V_{out}}{V_{in}} = \frac{j2\pi f t_c}{1 + j2\pi f t_c} \qquad\qquad (2.97)$$

If $t_c \ll 1$, then $H(f) \approx j2\pi f t_c$. ⭐ SIMULATION **HighPassFilter:** For further details.

Making use of the Fourier transform property that $\frac{ds(t)}{dt} \longleftrightarrow (j2\pi f)\,S(f)$, the output $y(t)$ of the high pass filter, which is given by the inverse FT of $H(f)\,S(f) = j2\pi f t_c S(f)$, is simply

$$
\begin{aligned}
y(t) &= t_c \frac{ds(t)}{dt} \qquad\qquad\qquad\qquad\qquad\qquad (2.98)\\
&= t_c \frac{d\left[A_c \cos\left(2\pi f_c t + \phi(t)\right)\right]}{dt}\\
&= -t_c A_c \left[2\pi f_c + \frac{d\phi(t)}{dt}\right] \sin\left(2\pi f_c t + \phi(t)\right)
\end{aligned}
$$

and thus the envelope detector output $y_{envelope}(t)$ is given by

$$
\begin{aligned}
y_{envelope}(t) &= t_c A_c \left[2\pi f_c + \frac{d\phi(t)}{dt}\right] = t_c A_c \left[2\pi f_c + k_f m(t)\right] \quad (2.99)\\
&= t_c A_c \left(2\pi f_c\right) + t_c A_c k_f m(t)
\end{aligned}
$$

where $\frac{d\phi(t)}{dt} = k_f m(t)$ for an FM signal. We have to first remove the d.c. term $t_c A_c (2\pi f_c)$ and divide $y_{envelope}(t)$ by a factor of $t_c A_c k_f$ to recover $m(t)$ as confirmed in ⭐ SIMULATION **P26**.

(b) ⭐ SIMULATION **P26:** For the solution to part(b).

Chapter 3

Digital Communications

3.1 Solutions

1. ▶VIDEO SOLUTION **P1:** Let $N_{zero}=$ Number of zero '0' binary digits
Let $N_{one}=$ Number of one '1' binary digits

(a) Probability of a binary digit zero

$$P_{zero} = \frac{N_{zero}}{N_{bits}} = \frac{13}{25} = 0.52 \qquad (3.1)$$

Similarly now find P_{one}.

(b) Probability of a binary digit one

$$P_{one} = \frac{N_{one}}{N_{bits}} = \frac{12}{25} = 0.48 \qquad (3.2)$$

Hence $P_{zero} + P_{one} = 0.52 + 0.48 = 1$ as expected.

2. ▶VIDEO SOLUTION **P2:** (a) Simply

$$
\begin{aligned}
P(1001|1001) &= P(1|1)P(0|0)P(0|0)P(1|1) \qquad (3.3)\\
&= (1-a)^4 = 0.6561
\end{aligned}
$$

(b) Similarly

$$P(0000|1001) = P(0|1)P(0|0)P(0|0)P(0|1) \qquad (3.4)$$
$$= (1-\alpha)^2 \alpha^2 = 0.0081$$

(c) Let "0" represent no error and let "1" represent an error. For one error in any position, the possible combinations are

Combination   Probability

Combination	Probability
0001	$(1-\alpha)^3 \alpha$
0010	$(1-\alpha)^3 \alpha$
0100	$(1-\alpha)^3 \alpha$
1000	$(1-\alpha)^3 \alpha$

Hence probability of one error in any position = Probability of error combination 0001 OR 0010 OR 0100 OR 1000

$$= (1-\alpha)^3 \alpha + (1-\alpha)^3 \alpha + (1-\alpha)^3 \alpha + (1-\alpha)^3 \alpha \qquad (3.5)$$
$$= 4(1-\alpha)^3 \alpha = 0.29$$

Equivalently, the probability of 1 error

$$P(n) = \binom{4}{1}(1-\alpha)^3 \alpha \qquad (3.6)$$
$$= \frac{4!}{(4-1)!1!}(1-\alpha)^3 \alpha$$
$$= \frac{4*3*2*1}{3*2*1*1}(1-\alpha)^3 \alpha$$
$$= 4(1-\alpha)^3 \alpha = 0.29$$

3. ▷ VIDEO SOLUTION **P3:** (a) From the previous question, we may now simply state that

Probability of n errors

$$P(n) = \binom{N}{n} (1-\alpha)^{N-n} \alpha^n \qquad (3.7)$$

(b) Probability of n errors or less,

$$P_{less}(n) = \binom{N}{n} (1-\alpha)^{N-n} \alpha^n \qquad (3.8)$$

OR

$$\binom{N}{n-1} (1-\alpha)^{N-n-1} \alpha^{n-1} \qquad (3.9)$$

OR

$$\binom{N}{0} (1-\alpha)^{N} \alpha^0 \qquad (3.10)$$

Hence

$$P_{less}(n) = \sum_{i=0}^{n} \binom{N}{i} (1-\alpha)^{N-i} \alpha^i \qquad (3.11)$$

4. ▶ VIDEO SOLUTION **P4:** Since

$$P(y_2|x_1) + P(y_1|x_1) = 1 \qquad (3.12)$$

$$P(y_1|x_1) = 1 - P(y_2|x_1) = 1 - 0.1 = 0.9 \qquad (3.13)$$

Similarly,

$$P(y_2|x_2) + P(y_1|x_2) = 1 \qquad (3.14)$$

$$P(y_2|x_2) = 1 - P(y_1|x_2) = 1 - 0.2 = 0.8 \qquad (3.15)$$

Also

$$P(x_1) + P(x_2) = 1 \tag{3.16}$$

hence

$$\begin{aligned} P(x_2) &= 1 - P(x_1) \\ &= 1 - 0.3 = 0.7 \end{aligned} \tag{3.17}$$

(a)

$$\begin{aligned} P(y_1) &= P(y_1|x_1)P(x_1) + P(y_1|x_2)P(x_2) \\ &= (0.9 * 0.3) + (0.2 * 0.7) = 0.41 \end{aligned} \tag{3.18}$$

(b)

$$\begin{aligned} P(y_2) &= P(y_2|x_1)P(x_1) + P(y_2|x_2)P(x_2) \\ &= (0.1 * 0.3) + (0.8 * 0.7) = 0.59 \end{aligned} \tag{3.19}$$

As expected, $P(y_1) + P(y_2) = 0.41 + 0.59 = 1$.

(c)

$$P(y_1, x_1) = P(y_1|x_1)P(x_1) \tag{3.20}$$

and

$$P(x_1, y_1) = P(x_1|y_1)P(y_1) \tag{3.21}$$

Hence

$$P(x_1|y_1)P(y_1) = P(y_1|x_1)P(x_1) \tag{3.22}$$

$$P(x_1|y_1) = \frac{P(y_1|x_1)P(x_1)}{P(y_1)} = \frac{0.9 * 0.3}{0.41} = 0.66 \tag{3.23}$$

$$P(x_2|y_1) = \frac{P(y_1|x_2)P(x_2)}{P(y_1)} = \frac{0.2 * 0.7}{0.41} = 0.34 \tag{3.24}$$

$$P(x_1|y_2) = \frac{P(y_2|x_1)P(x_1)}{P(y_2)} = \frac{0.1 * 0.3}{0.59} = 0.05 \qquad (3.25)$$

$$P(x_2|y_2) = \frac{P(y_2|x_2)P(x_2)}{P(y_2)} = \frac{0.8 * 0.7}{0.59} = 0.95 \qquad (3.26)$$

As a check, we expect $P(x_1|y_1) + P(x_2|y_1) = 0.66 + 0.34 = 1$ and $P(x_1|y_2) + P(x_2|y_2) = 0.05 + 0.95 = 1$

5. This probability is simply

$$P(x_1|y_1) = \frac{P(y_1|x_1)P(x_1)}{P(y_1)} = 0.66 \qquad (3.27)$$

- ⭐ SIMULATION **P5:** For the full solution of the previous problem and this problem. Alter the number such that $P(y_1|x_2) = P(y_2|x_1)$ i.e. consider what happens in a BSC. What do you conclude ?

6. ▶ VIDEO SOLUTION **P6:** The probabilities are given by

$$P(y_1) = P(y_1|x_1)P(x_1) = pP(x_1) \qquad (3.28)$$

$$P(y_2) = P(y_2|x_1)P(x_1) + P(y_2|x_2)P(x_2) \qquad (3.29)$$

$$P(y_2) = (1-p)\,P(x_1) + (1-p)\,P(x_2) = (1-p)\,(P(x_1) + P(x_2)) = 1 - p \qquad (3.30)$$

$$P(y_3) = P(y_3|x_2)P(x_2) = pP(x_2) \qquad (3.31)$$

7. ▶ VIDEO SOLUTION **P7:** (a)

$$P(4) = P(4|0)P(0) + P(4|1)P(1) \qquad (3.32)$$

(b)

$$P(0|4) = \frac{P(4|0)P(0)}{P(4)} \tag{3.33}$$

where $P(4)$ is as given in part (a)

8. ▶VIDEO SOLUTION **P8:** (a) Suppose the all-zero code word is transmitted. The table below shows the minimum number of channel errors required for a decoding error.

n	Code word transmitted	Channel errors required for decoding error
3	000	2
5	00000	3
7	0000000	4

Thus in general, require $\left(\dfrac{n-1}{2}+1\right)$ errors or more.

(b) For i errors in any position within n code digits, there are $\dfrac{n!}{(n-i)!i!}$ combinations. For example, suppose a binary digit zero is transmitted using $n = 3$. Then for $i = 2$, there are $\dfrac{3!}{(3-2)!2!} = 3$ combinations as highlighted in the following table. The probability of each received noisy code word and the corresponding decoded bit are also shown.

Received noisy code word	Probability	Decoded bit
000	$(1-\alpha)^3$	0
001	$\alpha(1-\alpha)^2$	0
010	$\alpha(1-\alpha)^2$	0
011	$\alpha^2(1-\alpha)$	1
100	$\alpha(1-\alpha)^2$	0
101	$\alpha^2(1-\alpha)$	1
110	$\alpha^2(1-\alpha)$	1
111	α^3	1

Hence probability of two errors $= \alpha^2(1-\alpha) + \alpha^2(1-\alpha) + \alpha^2(1-\alpha) = 3\alpha^2(1-\alpha)$. Thus in general,

$$P_{rep}(n, \alpha) = \sum_{i=\frac{(n-1)}{2}+1}^{n} \frac{n!}{(n-i)!i!} \alpha^i (1-\alpha)^{n-i} \qquad (3.34)$$

9. ▶ VIDEO SOLUTION **P9:** For $n = 2$, all the possible combinations of binary words with n digits (i.e. 2^n tuple) are

(00)
(01)
(10)
(11)

Consider for example $(x_1, x_2) = (10)$. Then $y_1 = x_1 = 1$ and $y_2 = (x_2 \oplus x_1) = 1$.

• ⭐ SIMULATION **P9:** For the rest of the solution.

10. ▶ VIDEO SOLUTION **P10:**

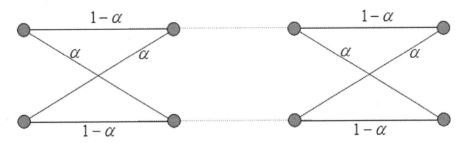

There is an error at the receiver only if there are an odd number of errors within the N BSC channels. The probability of n BSC independent errors out of N channels is given by

$$P(n) = \frac{N!}{(N-n)!n!} \alpha^n (1-\alpha)^{N-n} \qquad (3.35)$$

Hence for N BSC channels, the probability of a binary digit error P_e is thus given by

$$P_e = P(1) + P(3) + P(5) + ... + P(N) = \sum_{n=1,3,\cdots}^{N} P(n) \qquad (3.36)$$

⭐ SIMULATION **P10:** Illustrated that for $N\alpha \ll 1$, we find $P_e = N\alpha$.

Chapter 4

Information Theory

4.1 Solutions

1. ▶VIDEO SOLUTION **P1:** (a) Since $P_A + P_B + P_C = 1$, $P_B = P_C$ and $P_A = p$

$$P_B = \frac{1-p}{2} \tag{4.1}$$

Source entropy

$$
\begin{aligned}
H(X) &= \sum_{i=1}^{3} P(x_i) \log_2 \left(\frac{1}{P(x_i)} \right) \tag{4.2} \\
&= P_A \log_2 \left(\frac{1}{P_A} \right) + P_B \log_2 \left(\frac{1}{P_B} \right) + P_C \log_2 \left(\frac{1}{P_C} \right) \\
&= p \log_2 \left(\frac{1}{p} \right) + \left(\frac{1-p}{2} \right) \log_2 \left(\frac{2}{1-p} \right) + \left(\frac{1-p}{2} \right) \log_2 \left(\frac{2}{1-p} \right) \\
&= p \log_2 \left(\frac{1}{p} \right) + (1-p) \log_2 \left(\frac{2}{1-p} \right)
\end{aligned}
$$

Source entropy is a maximum when all the symbols are equally likely i.e. $P_A = P_B = P_C$, in which case $H(X) = \log_2 3$, as evident from the graph below.

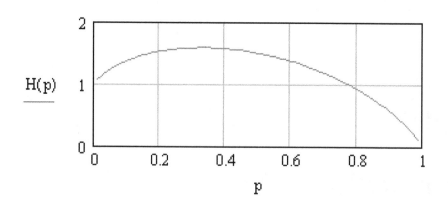

(b) For

$$P_A = \frac{1}{2}, P_B = P_C = \frac{1-0.5}{2} = \frac{1}{4} \tag{4.3}$$

$$P(CAABBAC) = \left(\frac{1}{4}\right)^4 \left(\frac{1}{2}\right)^3 = 0.000488 \tag{4.4}$$

Thus information content $= \log_2\left(\dfrac{1}{0.000488}\right) = \dfrac{\log_{10}\left(\dfrac{1}{0.000488}\right)}{\log_{10}(2)} = 11$ bits.

(c) Expected information content $= 7^*\, H(X)$, where

$$H(X) = p\log_2\left(\frac{1}{p}\right) + (1-p)\log_2\left(\frac{2}{1-p}\right) \tag{4.5}$$

For $p = 1/2$, $H(X) = 1.5$ bits per symbol. Hence expected information content $= 7^*1.5 = 10.5$ bits.

(d) Source entropy $H(X) = \log_2(3) = 1.585$bits per symbol.

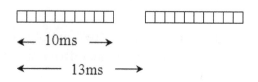

Notice that 10 symbols are produced every 13 ms. Hence the symbol rate $r = \dfrac{10}{13} * 1000 = 769$ symbols per second. Hence information rate $R = rH(X) = 769 * 1.585 = 1219$ bits per second.

- ⭐ **SIMULATION P1:** For further details.

2. ▶ **VIDEO SOLUTION P2:** (a) Let N_A = number of A's output in one second

Let N_B = number of B's output in one second

$$T_A N_A + T_B N_B = 1 \tag{4.6}$$

Now $P_A = \dfrac{N_A}{N}$ and $P_B = \dfrac{N_B}{N}$, where N is the total number of symbols output in one second.

Now since $T_A P_A N + T_B P_B N = 1$, we find

$$N = \frac{1}{T_A P_A + T_B P_B} \text{ symbols per second} \tag{4.7}$$

Hence information rate

$$R = \left(\frac{1}{T_A P_A + T_B P_B} \right) H(X) \tag{4.8}$$

where

$$H(X) = P_A \log_2 \left(\frac{1}{P_A} \right) + P_B \log_2 \left(\frac{1}{P_B} \right) \tag{4.9}$$

(b) Let symbol A correspond to a dot and let symbol B correspond to dash.

$$P_A = 0.7, T_A = 0.1, T_B = 0.2 \tag{4.10}$$

$$P_A + P_B = 1 P_B = 1 - P_A = 1 - 0.7 = 0.3 \tag{4.11}$$

Hence $R = 6.8$ bits/sec.

- ⭐ SIMULATION **P2:** For further details.

3. ▶ VIDEO SOLUTION **P3:** ⭐ SIMULATION **P3:** For the full solution.

- Experiment with the variables to understand and appreciate their influence on the answer.

4. ▶ VIDEO SOLUTION **P4:** ⭐ SIMULATION **P4:** For the full solution. Experiment with the variables to understand and appreciate their influence on the answer.

5. ▶ VIDEO SOLUTION **P5:** (a)

$$I_A = \log_2 \left(\frac{1}{1/2} \right) = \log_2(2) = 1 \tag{4.12}$$

Similarly,

$$I_B = 2, I_C = 3, I_D = 3 \tag{4.13}$$

(b) Source Code II

A [0]
B [10]
C [110]
D [111]
Encoding

BBACDBD = 1010011011110111

Decoding

1010011011110111

- First digit 1 is not a valid code word
- First two digits 10 is a valid code word = B
- First three digits 101 is not a valid code word
- Hence first symbol must be B
- Remaining to decode 10011011110111
- Repeat process until all the symbols are decoded.

(c)

Symbol	Probability	Source Code I	Self Information	Code Length
A	1/2	0	1	1
B	1/4	01	2	2
C	1/8	011	3	3
D	1/8	0111	3	4

Source entropy

$$H(X) = 1\left(\frac{1}{2}\right) + 2\left(\frac{1}{4}\right) + 3\left(\frac{1}{8}\right) + 3\left(\frac{1}{8}\right) \tag{4.14}$$

Average code length

$$\overline{L} = 1\left(\frac{1}{2}\right) + 2\left(\frac{1}{4}\right) + 3\left(\frac{1}{8}\right) + 4\left(\frac{1}{8}\right) \tag{4.15}$$

Efficiency

$$\eta = \frac{H(X)}{\overline{L}} * 100 = 93\% \tag{4.16}$$

Similarly, the efficiency of Code II = 100 % and code III = 87.5 %.

(d)

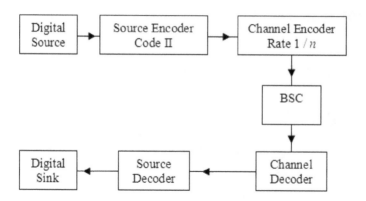

Using Shannon's channel coding theorem $R_{code}\Omega(p) \leq C_s$ where p is the probability with which a binary digit zero is output from the source encoder. Recall that the probability of a binary digit zero output from the source encoder

$$p_{zero} = \frac{\sum_{i=1}^{M} P(x_i) L_i^{(zero)}}{\overline{L}} \qquad (4.17)$$

Hence in this case,

$$
\begin{aligned}
p &= \frac{\sum_{i=1}^{4} P(x_i) L_i^{(zero)}}{\overline{L}} \qquad (4.18) \\
&= \frac{1\left(\frac{1}{2}\right) + 1\left(\frac{1}{4}\right) + 1\left(\frac{1}{8}\right) + 0\left(\frac{1}{8}\right)}{1\left(\frac{1}{2}\right) + 2\left(\frac{1}{4}\right) + 3\left(\frac{1}{8}\right) + 3\left(\frac{1}{8}\right)} \\
&= \frac{0.875}{1.75} = 0.5
\end{aligned}
$$

as expected since the efficiency of this code is 100 %. In this case, recall that $C_s = 1 - \Omega(\alpha)$ bits/symbol. Since $\Omega(p) = 1$, we find that Shannon's coding theorem requires $R_{code} \leq 1 - \Omega(\alpha)$. But $R_{code} = 1/2$

$$\therefore \frac{1}{2} \leq 1 - \Omega(\alpha) \tag{4.19}$$

or equivalently

$$\Omega(\alpha) \leq \frac{1}{2} \tag{4.20}$$

From the figure below, we find that $\Omega(0.11) = \frac{1}{2}$. Therefore require $\alpha \leq 0.11$ for reliable communication using a rate $\frac{1}{2}$ channel encoder and decoder.

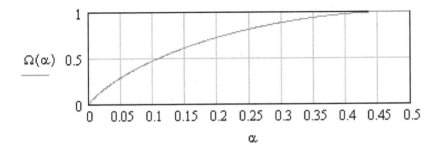

- ⭐ SIMULATION P5: To experiment with different source probabilities and codes.

6. ▶ VIDEO SOLUTION P6: Recall the average information transferred per symbol

$$I(X,Y) = \Omega(p + \alpha - 2\alpha p) - \Omega(\alpha) \tag{4.21}$$

Its variation with α for certain values of p are shown in the next part of this solution. Notice that no information is transferred if $\alpha = 0.5$ for any value of p. Also, of course no information is transferred if $p = 0$ or 1 for any value of α.

$$I(\alpha,p) := \Omega(p + \alpha - 2 \cdot \alpha \cdot p) - \Omega(\alpha)$$

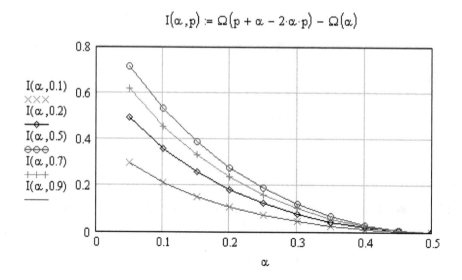

- SIMULATION **P6:** for further details.

7. VIDEO SOLUTION **P7:** (a) The DMC is a BSC if $P(0|1) = P(1|0)$.

(b) Given $P(x_1) = p$, $P(y_1|x_2) = P(0|1) = \mu$, $P(y_2|x_1) = P(1|0) = \lambda$

$$P(x_2) = 1 - p \tag{4.22}$$

$$P(y_2|x_2) = P(1|1) = 1 - \mu \tag{4.23}$$

$$P(y_1|x_1) = P(0|0) = 1 - \lambda \tag{4.24}$$

$$P(y_1) = P(y_1|x_1)P(x_1) + P(y_1|x_2)P(x_2)$$

$$\begin{aligned} P(y_1) &= (1 - \lambda)\,p + \mu\,(1 - p) \\ &= p - \lambda p + \mu - \mu p = \mu + p(1 - \lambda - \mu) \end{aligned} \tag{4.25}$$

Entropy

$$H(Y) = P(y_1) \log_2\left(\frac{1}{P(y_1)}\right) + P(y_2) \log_2\left(\frac{1}{P(y_2)}\right) \tag{4.26}$$

$$\begin{aligned} &= P(y_1) \log_2\left(\frac{1}{P(y_1)}\right) + (1 - P(y_1)) \log_2\left(\frac{1}{1 - P(y_1)}\right) \quad (4.27) \\ &= \Omega\left(P(y_1)\right) \end{aligned}$$

$$H(Y) = \Omega\left(\mu + p\left(1 - \lambda - \mu\right)\right) \tag{4.28}$$

$$H(Y|X) = P(y_1|x_1)P(x_1) \log_2\left(\frac{1}{P(y_1|x_1)}\right) \tag{4.29}$$

$$+ P(y_1|x_2)P(x_2) \log_2\left(\frac{1}{P(y_1|x_2)}\right) \tag{4.30}$$

$$+ P(y_2|x_1)P(x_1) \log_2\left(\frac{1}{P(y_2|x_1)}\right)$$

$$+ P(y_2|x_2)P(x_2) \log_2\left(\frac{1}{P(y_2|x_2)}\right)$$

$$\begin{aligned} H(Y|X) &= (1 - \lambda)\, p \log_2\left(\frac{1}{1 - \lambda}\right) + \mu\,(1 - p) \log_2\left(\frac{1}{\mu}\right) \quad (4.31) \\ &\quad + \lambda p \log_2\left(\frac{1}{\lambda}\right) + (1 - \mu)\,(1 - p) \log_2\left(\frac{1}{1 - \mu}\right) \end{aligned}$$

$$\begin{aligned} H(Y|X) &= (1 - \lambda)\, p \log_2\left(\frac{1}{1 - \lambda}\right) + \mu\,(1 - p) \log_2\left(\frac{1}{\mu}\right) \\ &\quad + \lambda p \log_2\left(\frac{1}{\lambda}\right) + (1 - \mu)\,(1 - p) \log_2\left(\frac{1}{1 - \mu}\right) \\ &= p\Omega(\lambda) + (1 - p)\,\Omega(\mu) \end{aligned}$$

Hence

$$I(X,Y) = \Omega\left(\mu + p\left(1 - \lambda - \mu\right)\right) - p\Omega(\lambda) - (1 - p)\,\Omega(\mu) \qquad (4.32)$$

$$P_e = P(y_2|x_1)P(x_1) + P(y_1|x_2)P(x_2) \qquad (4.33)$$

$$P_e = \lambda p + \mu(1 - p) \qquad (4.34)$$

(c) $I(X,Y) = 0.397$ and $P_e = 0.15$

(d) By inspection of the forward transition probabilities, we simply require $\mu = 1 - \lambda$. To verify this,

$$
\begin{aligned}
I(X,Y) &= \Omega\left(\mu + p\left(1 - \lambda - \mu\right)\right) - p\Omega(\lambda) - (1 - p)\,\Omega(\mu) \qquad (4.35)\\
&= \Omega\left(\mu\right) - p\Omega(\lambda) - (1 - p)\,\Omega(\mu)\\
&= \Omega\left(\mu\right) - p\Omega(\lambda) - \Omega(\mu) + p\Omega(\mu) = 0
\end{aligned}
$$

For $\mu = 1 - \lambda$ and $p = 0.5$ we find that (as expected)

$$
\begin{aligned}
P_e &= \lambda p + \mu(1 - p) \qquad (4.36)\\
&= \lambda p + (1 - \lambda)(1 - p) = \lambda p + (1 - p - \lambda + \lambda p)\\
&= 1 - p - \lambda + 2\lambda p = 1 - 0.5 - \lambda + \lambda = 0.5
\end{aligned}
$$

(e) For $\lambda = \mu$

$$
\begin{aligned}
I(X,Y) &= \Omega\left(\mu + p\left(1 - \lambda - \mu\right)\right) - p\Omega(\lambda) - (1 - p)\,\Omega(\mu) \qquad (4.37)\\
&= \Omega\left(\lambda + p\left(1 - 2\lambda\right)\right) - p\Omega(\lambda) - (1 - p)\,\Omega(\lambda)\\
&= \Omega\left(\lambda + p\left(1 - 2\lambda\right)\right) - p\Omega(\lambda) - (1 - p)\,\Omega(\lambda)\\
&= \Omega\left(\lambda + p - 2p\lambda\right) - p\Omega(\lambda) - \Omega(\lambda) + p\Omega(\lambda)\\
&= \Omega\left(\lambda + p - 2p\lambda\right) - \Omega(\lambda)
\end{aligned}
$$

This expression is a maximum when $p = 0.5$. Thus the channel capacity $= 1 - \Omega(\lambda)$, as expected.

(f) From the graph below, we estimate $p = 0.52$.

$$I(\mu, \lambda, p) := \Omega\big[\mu + (1 - \lambda - \mu)\cdot p\big] - \Omega(\lambda)\cdot p - (1 - p)\cdot\Omega(\mu)$$

- ⭐ SIMULATION **P7:** For further details. Experiment with the variables to understand and appreciate their influence on the answer.

8. ▶️VIDEO SOLUTION **P8:** (a)

$$S_i \in \{1, 2, 3, 4, 5, 6\} \qquad P(s_i) = \frac{1}{6}$$

$$p_{zero} = \frac{\sum_{i=1}^{M} P(x_i) L_i^{(zero)}}{\bar{L}} \tag{4.38}$$

$$= \frac{\frac{1}{6}(2 + 2 + 1 + 2 + 1 + 1)}{3} = \frac{9}{6*3} = \frac{1}{2} \tag{4.39}$$

(b)

$$H(X) = \sum_{i=1}^{6} P(s_i) \log_2 \left(\frac{1}{P(s_i)} \right) \tag{4.40}$$

$$= \sum_{i=1}^{6} \frac{1}{6} \log_2 (6) = \log_2 (6) = 2.585 \text{ bits/symbol}$$

(c) Information rate $R = r_s H(X) = 1 * 2.585 = 2.585$ bits/sec.

(d) By inspection, $\overline{L} = 3$. Hence the binary digits/sec output from the source encoder, $r_b = r_s \overline{L} = 3$ binary digits/sec.

Recall Shannon's source coding theorem may be stated as $r_b \geq R$. Hence $r_b \geq 2.585$ binary digits/sec. Efficiency of the source encoder $= \frac{H(X)}{\overline{L}} *$
$100 = \frac{2.585}{3} * 100 = 86\%$

(e) Label the transition diagram

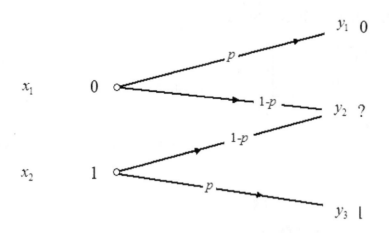

$$P(y_1) = P(y_1|x_1)P(x_1) = p * 1/2 = \frac{p}{2} = P(y_3) \tag{4.41}$$

$$P(y_2) = P(y_2|x_1)P(x_1) + P(y_2|x_2)P(x_2) \tag{4.42}$$

$$P(y_2) = (1-p)\left(\frac{1}{2}\right) + (1-p)\left(\frac{1}{2}\right) = (1-p) \tag{4.43}$$

$$
\begin{aligned}
H(Y) &= P(y_1)\log_2\left(\frac{1}{P(y_1)}\right) + P(y_2)\log_2\left(\frac{1}{P(y_2)}\right) \\
&\quad + P(y_3)\log_2\left(\frac{1}{P(y_3)}\right) \tag{4.44} \\
&= 2\left(\frac{p}{2}\log_2\left(\frac{2}{p}\right)\right) + (1-p)\log_2\left(\frac{1}{1-p}\right) = \Omega(p) + p
\end{aligned}
$$

$$
\begin{aligned}
&H(Y|X) \tag{4.45} \\
&= P(x_1)\left[\begin{array}{l}P(y_1|x_1)\log_2\left(\frac{1}{P(y_1|x_1)}\right) + P(y_2|x_1)\log_2\left(\frac{1}{P(y_2|x_1)}\right) \\ +P(y_3|x_1)\log_2\left(\frac{1}{P(y_3|x_1)}\right)\end{array}\right] \\
&\quad + P(x_2)\left[\begin{array}{l}P(y_1|x_2)\log_2\left(\frac{1}{P(y_1|x_2)}\right) + P(y_2|x_2)\log_2\left(\frac{1}{P(y_2|x_2)}\right) \\ +P(y_3|x_2)\log_2\left(\frac{1}{P(y_3|x_2)}\right)\end{array}\right]
\end{aligned}
$$

$$H(Y|X) = \frac{1}{2}\left(\Omega(p) + \Omega(p)\right) = \Omega(p) \tag{4.46}$$

Therefore, average mutual information

$$I(X,Y) = H(Y) - H(Y|X) = \Omega(p) + p - \Omega(p) = p \text{ bits/symbol} \tag{4.47}$$

(f) Maximum information rate through the channel $= I(X,Y)r_b = 3p$ bits/sec.

(g) If the die was biased, then change the source encoder code to still ensure that the probability of a binary digit zero $= 0.5$.

9. 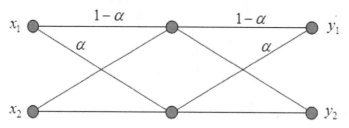VIDEO SOLUTION **P9**:

$$P(y_1|x_1) = P(y_2|x_2) = (1 - \alpha)^2 + \alpha^2 \tag{4.48}$$

$$P(y_1|x_2) = \alpha(1 - \alpha) + (1 - \alpha)\alpha = 2\alpha(1 - \alpha) \tag{4.49}$$

As a check, note that

$$
\begin{aligned}
P(y_1|x_2) &+ P(y_2|x_2) \\
&= (1 - \alpha)^2 + \alpha^2 + 2\alpha(1 - \alpha) \\
&= 1 - 2\alpha + \alpha^2 + \alpha^2 + 2\alpha - 2\alpha^2 = 1
\end{aligned}
\tag{4.50}
$$

as expected. The two BSCs can be modeled as a single BSC, in which the crossover probability $\beta = P(y_1|x_2) = 2\alpha(1 - \alpha)$, as shown below.

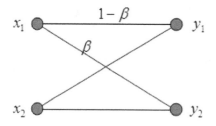

Recall that the average mutual information

$$I(X, Y) = \Omega(p + \beta - 2\beta p) - \Omega(\beta) \tag{4.51}$$

For $p = 0.5$, the channel capacity

$$
\begin{aligned}
C_s &= \Omega\left(0.5\right) - \Omega\left(\beta\right) \\
&= 1 - \Omega\left(\beta\right) = 1 - \Omega\left(2\alpha(1-\alpha)\right)
\end{aligned}
\tag{4.52}
$$

(c) Recall that for reliable communication over a DMC, Shannon's channel coding theorem requires that

$$
R_{code}\Omega(p) \le C_s
\tag{4.53}
$$

For $p = 0.5$, $\Omega(0.5) = 1$. Thus $R_{code} \le C_s$. For $R_{code} = 1/2$,

$$
0.5 \le 1 - \Omega\left(\beta\right)
\tag{4.54}
$$

Thus we require $\Omega\left(\beta\right) \le 0.5$. From the graph shown in the next part of this solution, we note that this condition is satisfied for approximately $\beta \le 0.1$.

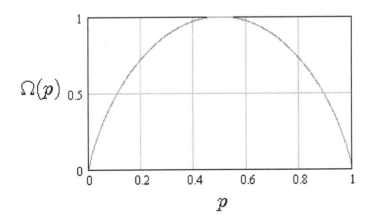

Since $\beta = 2\alpha(1 - \alpha)$, we require

$$
2\alpha(1 - \alpha) \le 0.1
\tag{4.55}
$$

or equivalently,

$$
2\alpha^2 - 2\alpha + 0.1 \ge 0
\tag{4.56}
$$

Let $y(\alpha) = 2\alpha^2 - 2\alpha + 0.1$. A plot of this polynomial is shown below.

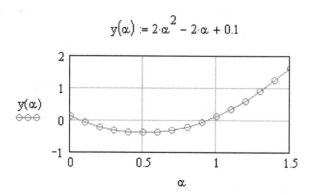

Using the standard formula to solve the quadratic equation $2\alpha^2 - 2\alpha + 0.1 = 0$, we find the two possible solutions to be 0.053 and 0.947, which correspond to the intersection of the curve $y(\alpha)$ and the horizontal line for $y = 0$. Thus the channel coding theorem is satisfied for $\alpha \leq 0.053$, since $\alpha = 0.5$ corresponds to maximum noise.

10. ▶VIDEO SOLUTION **P10:**

$$P_* = \frac{10}{15} = \frac{2}{3}$$

$$P_= = \frac{5}{15} = \frac{1}{3}$$

$$P(*| =) = \frac{2}{5}$$

$$P(= |*) = \frac{4}{10} = \frac{2}{5}$$

$$P(*|*) = \frac{3}{5}$$

$$P(= | =) = \frac{3}{5}$$

(b) The average information transferred per symbol

$$I(X,Y) = \Omega\,(p + \alpha - 2\alpha p) - \Omega\,(\alpha) \tag{4.57}$$

where p was the probability of a binary digit zero entering the BSC and α is the BSC crossover probability. In this case, we may set $p = P_*$. For $\alpha = \frac{2}{5}$, we find that $I(X,Y) = 0.026$ bits/symbol. If we had taken $p = P_=$, the answer is still the same. To determine the average number of symbols per sec:

Let N = number of symbols output in T seconds. The number of symbols per second output on average $= r_s = \dfrac{N}{T}$.

Let N_* = number of * symbols within this output stream
Let $N_=$ = number of = symbols within this output stream
Let T_* = Time duration of the symbol *
Let $T_=$ = Time duration of the symbol =
Let T_{gap} = Time duration of the gap between each symbol

Then

$$T = N_* T_* + N_= T_= + (N_* + N_= - 1)\,T_{gap} \tag{4.58}$$

$$\frac{T}{N} = \frac{N_*}{N}T_* + \frac{N_=}{N}T_= + \frac{(N_* + N_= - 1)}{N}T_{gap} \tag{4.59}$$

Thus

$$\frac{1}{r_s} = P_* T_* + P_= T_= + \left(P_* + P_= - \frac{1}{N}\right)T_{gap} \tag{4.60}$$

but for large N, $\dfrac{1}{N} \approx 0$.

Hence

$$r_s = \frac{1}{P_* T_* + P_= T_= + (P_* + P_=)\,T_{gap}} = 37.5 \text{ symbols/sec} \tag{4.61}$$

Finally the information rate $= 37.5 * 0.026 = 0.975$ bits/sec.

- ⭐ SIMULATION **P10:** For further details.   Be sure to experiment with the variables to understand and appreciate their influence on the answer.

11. ▷VIDEO SOLUTION **P11** and ⭐ SIMULATION **P11:** For the full solution.

12. ▷VIDEO SOLUTION **P12:**   The source entropy in this case is given by $H(X) = \log_2 M$. Shannon's source coding theorem states that the average code length $\overline{L} \geq H(X)$. Hence, for optimum source coding, we want $\overline{L} = \log_2 M$.

$$\overline{L} = \sum_{i=1}^{M} P(x_i)L_i = \frac{1}{M} \sum_{i=1}^{M} L_i = \log_2 M \tag{4.62}$$

Thus, for a fixed length code, let $L_i = l$.

$$\frac{1}{M}(Ml) = \log_2 M \tag{4.63}$$

from which we find that $l = \log_2 M$.

13. ▷VIDEO SOLUTION **P13** and ⭐SIMULATION **P13:** For the full solution.   You may use this code to solve many other similar problems.   Simply change the contents of the "sym" array to reflect the source probabilities.   If there are N symbols, then the array "sym" will contain N probabilities.

14. ▷VIDEO SOLUTION **P14** and ⭐SIMULATION **P14:** For the full solution.

15. ▷VIDEO SOLUTION **P15:**

0	1	11	110	111	01	1111	1101
1	2	3	4	5	6	7	8
		2	3	3	1	5	4
		0101	0110	0111	0011	1011	1001

Number of input digits = 18
Number of output digits = 24
Compression factor = $24/18 = 1.3$. This factor will only be less than one for large input sequences.

- ⭐ SIMULATION **P15:** For further details. Experiment with the variables to understand and appreciate their influence on the answer.

16. ▶ VIDEO SOLUTION **P16** and ⭐ SIMULATION **P16:** For the full solution.

Chapter 5

Analog to Digital

5.1 Solutions

1. The Fourier Transform

$$S(f) = \begin{cases} 1 & if & |f| \leq \dfrac{A}{2\pi} \\ 0 & otherwise \end{cases} \qquad (5.1)$$

Thus $s(t)$ is a band-limited signal with $f_{\max} = \dfrac{A}{2\pi} = \dfrac{2\pi}{2\pi} = 1$ Hz. Thus $f_s = 2$ Hz.

- ⭐ SIMULATION **P1:** For the rest of the solution.

- Experiment with under sampling and over sampling. Change the expression for the input signal $s(t)$ to solve for yourself, numerous other similar problems.

2. ⭐ SIMULATION **P2:** For the full solution. Experiment with the variables to understand and appreciate their influence on the answer.

3. The Nyquist rate for this sinusoidal signal is obviously $2f_o$. By sampling at a frequency under $2f_o$, aliasing components are introduced. Specifically, we get an unwanted component at the frequency

$$(f_s - f_o) = 1.5f_o - f_o = \frac{f_o}{2}$$ (5.2)

Given that

$$f_c = \frac{f_s}{2} = \frac{3f_o}{2 * 2} = \frac{3f_o}{4}$$ (5.3)

the reconstructed signal will be

$$s(t) = A \cos\left(2\pi \frac{f_o}{2}t\right) = A \cos\left(\pi f_o t\right)$$ (5.4)

4. ⭐ SIMULATION **P4:** For the full solution.

5. (a) Given

$$B_{PCM} \geqslant \frac{r_b}{2}$$ (5.5)

$$
\begin{aligned}
r_b &= f_s \log_2 Q_{level} \\
&= 8000 \log_2 256 = 8000 * 8 = 64000
\end{aligned}
$$ (5.6)

Thus

$$B_{PCM} \geqslant 32000 Hz$$ (5.7)

Voice analog channel bandwidth $B_{analog} = 4$ kHz

Thus

$$\frac{B_{PCM}}{B_{analog}} \geqslant \frac{32000}{4000} = 8$$ (5.8)

or equivalently

$$B_{PCM} \geqslant 8B_{analog} \tag{5.9}$$

(b) At the Nyquist sampling rate, $f_s = 2B_{analog}$. The PCM first null signal bandwidth $B_{null} = r_b$.

$$\frac{B_{Null}}{B_{analog}} = \frac{r_b}{f_s/2} \tag{5.10}$$

$$\begin{aligned} \frac{B_{Null}}{B_{analog}} &= \frac{f_s \log_2(Q_{level})}{f_s/2} \\ &= 2\log_2(Q_{level}) \end{aligned} \tag{5.11}$$

6. (a) Given that $r_b = f_s \log_2(Q_{level})$, we require

$$f_s \log_2(Q_{level}) \leq 56000 \tag{5.12}$$

where $f_s \geq 2(3800)$. Therefore $\log_2(Q_{level}) \leq \dfrac{56000}{2\,(3800)}$ or equivalently

$$\log_2(Q_{level}) \leq 7.368 \tag{5.13}$$

Hence we set $\log_2(Q_{level}) = 7$, from which $Q_{level} = 2^7 = 128$ levels. Finally,

$$f_s = \frac{r_b}{\log_2(Q_{level})} = \frac{56000}{7} = 8000 \text{ samples/second} \tag{5.14}$$

(b) $T_b = \dfrac{1}{r_b} = \dfrac{1}{56000}$ seconds.

(c) First null bandwidth $= \dfrac{1}{T_b} = r_b = 56$ kHz. The absolute minimum bandwidth

$$B_{PCM}^{(min)} = \frac{f_s}{2}\log_2(Q_{level}) = \frac{r_b}{2} = \frac{56000}{2} = 28kHz \tag{5.15}$$

(d) Assuming a uniform PDF (i.e. $p(x) = \frac{1}{2A}$), the average $SQNR = Q^2_{level} = (128)^2$, which expressed in decibels is equal to

$$SQNR(dB) = 10 \log_{10} \left((128)^2\right) = 42.1 dB \tag{5.16}$$

● ⭐ SIMULATION **P6:** For further details.   Experiment with the variables to understand and appreciate their influence on the answer.

7. Consider first a uniform quantizer, for which we recall that

$$SQNR(dB) = 10 \log_{10} \left(\frac{3P_s}{A^2}\right) + 6k \tag{5.17}$$

For an increase in k by 1, the improvement in $SQNR(dB)$ is given by

$$10 \log_{10} \left(\frac{3P_s}{A^2}\right) + 6\,(k+1) - \left[10 \log_{10} \left(\frac{3P_s}{A^2}\right) + 6k\right] = 6 \text{ dB} \tag{5.18}$$

Similarly, for a non-uniform quantizer, recall that

$$SQNR(dB) = 10 \log_{10} \left(\frac{3}{[\ln{(1+\mu)}]^2}\right) + 6k \tag{5.19}$$

Once again, the the improvement in $SQNR(dB)$ is simply given by

$$10 \log_{10} \left(\frac{3}{[\ln{(1+\mu)}]^2}\right) + 6\,(k+1) \tag{5.20}$$
$$- \left[10 \log_{10} \left(\frac{3}{[\ln{(1+\mu)}]^2}\right) + 6k\right]$$
$$= \quad 6 dB$$

In general, given that $SQNR$ depends on Q^2_{level},

$$SQNR(dB) = 10\log_{10}\left(Q_{level}\right)^2 + constant \qquad (5.21)$$
$$= 20\log_{10}\left(Q_{level}\right) + constant$$

Given that $Q_{level} = 2^k$,

$$\tag{5.22}$$
$$SQNR(dB) = 20\log_{10}\left(2^k\right) + constant = 20k\log_{10}2 + constant$$

Therefore, the improvement in the *SQNR* is given by

$$20\left(k+1\right)\log_{10}2 - 20k\log_{10}2 = 20\log_{10}2\left(k+1-k\right) = 6\ \mathrm{dB} \qquad (5.23)$$

8. With stereo, we have two channels in which the sampling frequency is 44.1 kHz and 16 binary digits are used per PCM word. Thus

$$r_b = f_s\log_2 Q_{level} = 44100(16)(2) * 2 \qquad (5.24)$$
$$= 1411200\ \mathrm{binarydigits/sec}$$
$$= 176400\ \mathrm{bytes/second}$$

For a 700 MB CD-R, we can store $\dfrac{700(10)^6(8)}{1411200}$ =3968 seconds, or equivalently 66 minutes worth of music. With formatting of the CD-R, less storage space is available. Thus typically, about 1 hour's worth of CD quality can be stored on a 700 MB CD-R.

- ⭐ SIMULATION **P8**: For parts (a) and (b).

9. ⭐ SIMULATION **P9**: For the full solution.

10. ⭐ SIMULATION **P10:** For the full solution.   For part (b), we find the *SQNR* saturates to approx. 24 dB. One solution is to employ error-control coding to reduce P_e.

(c) Given that

$$SQNR = \frac{Q_{level}^2}{1 + 4\left(Q_{level}^2 - 1\right)P_e} \qquad (5.25)$$

as $P_e \to 0$,

$$SQNR \to Q_{level}^2 \qquad (5.26)$$

i.e. the SQNR is due only to quantizing errors.

11. The number binary digits per PCM word $n = \log_2 Q_{level}$. To ensure no reduction in the information throughput, these binary digits must be transmitted within $\dfrac{1}{f_s}$ seconds. i.e.

$$nT_b = \frac{1}{f_s} \qquad (5.27)$$

$$T_b = \frac{1}{f_s \log_2 Q_{level}} \qquad (5.28)$$

12. (a) We require

$$|e| \le \left(\frac{p}{100}\right)(2A) \qquad (5.29)$$

where $|e|_{max} = \dfrac{x}{2}$. i.e. we want $\dfrac{x}{2} \le \left(\dfrac{p}{100}\right)(2A)$. Since $x = \dfrac{2A}{Q_{level}}$, we find

$$\frac{A}{Q_{level}} \le \left(\frac{p}{100}\right)(2A) \qquad (5.30)$$

Thus $\dfrac{50}{p} \le Q_{level}$ or equivalently

$$Q_{level} \geq \frac{50}{p} \tag{5.31}$$

(b) $Q_{level} \geq \dfrac{50}{2}$ and so $2^n \geq 25$ from which

$$n \geq \frac{\log(25)}{\log(2)} \tag{5.32}$$
$$\geq 4.6$$

Thus we choose $n = 5$ binary digits.

13. (a) ⭐ SIMULATION **P13:** For the solution to part(a).

(b) Recall

$$|H(f)|^2 = \frac{PSD_{out}(f)}{PSD_{in}(f)} \tag{5.33}$$

For $|H(f)|^2 \, (dB) = -40$ dB or equivalently,

$$|H(f)|^2 = 10^{-4}, f = f_c \left(\frac{1}{10^{-4}} - 1\right)^{1/2n} = 5335 \text{ Hz} \tag{5.34}$$

Thus the sampling frequency required is $2(5335) = 10670$ Hz.

14. ⭐ SIMULATION **P14:** For the full solution.   Experiment with the variables to understand and appreciate their influence on the answer.

15.   Since the total area under the PDF is equal to 1, we find $Ah +$ $\dfrac{A}{2}\left(\dfrac{h}{2}\right)2 = 1$, from which $h = \dfrac{2}{3A}$. By symmetry, the mean value $x_m = 0$. Thus

$$\sigma_m^2 = \int\limits_{-\infty}^{\infty} (x - x_m)^2\, p(x)dx = \int\limits_{-A}^{A} x^2 p(x)dx \qquad (5.35)$$

$$
\begin{aligned}
\sigma_m^2 &= \int\limits_{-A}^{-A/2} \frac{1}{3A} x^2 dx + \int\limits_{-A/2}^{A/2} \frac{2}{3A} x^2 dx + \int\limits_{A/2}^{A} \frac{1}{3A} x^2 dx \qquad (5.36) \\[2ex]
&= \frac{2}{3A} \int\limits_{A/2}^{A} x^2 dx + \frac{2}{3A} \int\limits_{-A/2}^{A/2} x^2 dx = \frac{2}{3A} \left[\frac{x^3}{3} \right]_{A/2}^{A} + \frac{2}{3A} \left[\frac{x^3}{3} \right]_{-A/2}^{A/2} \\[2ex]
&= \sigma_m^2 = \frac{2}{3A} \left[\frac{A^3}{3} - \frac{A^3}{24} \right] + \frac{2}{3A} \left[\frac{2A^3}{24} \right] \\[2ex]
&= \frac{A^2}{4}
\end{aligned}
$$

The average quantization noise power or equiavalently

$$x_{mse} = \int\limits_{-\infty}^{\infty} (x - y)^2\, p(x)dx \qquad (5.37)$$

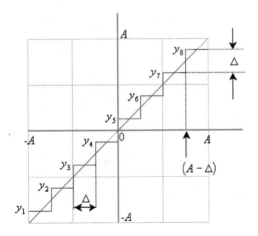

$$x_{mse} = \int_{-\infty}^{\infty} (x - y)^2 p(x) dx \tag{5.38}$$

$$= \int_{-A}^{-A+\Delta} (x - y_1)^2 p(x) dx + \int_{-A+\Delta}^{-A+2\Delta} (x - y_2)^2 p(x) dx +$$

$$\cdots + \int_{A-\Delta}^{A} (x - y_8)^2 p(x) dx$$

Let

$$N_1 = \int_{-A}^{-A+\Delta} (x - y_1)^2 p(x) dx, N_2 \tag{5.39}$$

$$= \int_{-A+\Delta}^{-A+2\Delta} (x - y_2)^2 p(x) dx$$

etc. so that

$$x_{mse} = N_1 + N_2 + \cdots + N_{Q_{level}} \tag{5.40}$$

By symmetry, $N_1 = N_2 = N_7 = N_8$ and $N_3 = N_4 = N_5 = N_6$.

$$N_8 = \int_{A-\Delta}^{A} (x - y_8)^2 \, p(x) dx = \frac{1}{3A} \int_{A-\Delta}^{A} \left(x - \left(A - \frac{\Delta}{2} \right) \right)^2 dx \qquad (5.41)$$

Now

$$N_5 = \int_{0}^{\Delta} (x - y_5)^2 \, p(x) dx = \frac{2}{3A} \int_{0}^{\Delta} \left(x - \left(\frac{\Delta}{2} \right) \right)^2 dx \qquad (5.42)$$

Let $u = x - A + \dfrac{\Delta}{2}$. Then $\dfrac{du}{dx} = 1$. Changing the integration limits, we find

$$N_8 \; = \; \frac{1}{3A} \int_{-\Delta/2}^{\Delta/2} u^2 du \qquad (5.43)$$

$$= \; \frac{1}{3A} \left[\frac{u^3}{3} \right]_{-\Delta/2}^{\Delta/2} = \frac{1}{3A} \left(\frac{2\Delta^3}{24} \right)$$

Since $Q_{level} = 8 = \dfrac{2A}{\Delta}$

$$N_8 = \frac{1}{3(4\Delta)} \left(\frac{2\Delta^3}{24} \right) = \frac{\Delta^2}{144} \qquad (5.44)$$

Let $u = x - \dfrac{\Delta}{2}$. Then $\dfrac{du}{dx} = 1$. Changing the integration limits, we find

$$N_5 \; = \; \frac{2}{3A} \int_{-\Delta/2}^{\Delta/2} u^2 du = \frac{2}{3A} \left[\frac{u^3}{3} \right]_{-\Delta/2}^{\Delta/2} \qquad (5.45)$$

$$= \; \frac{2}{3A} \left(\frac{2\Delta^3}{24} \right)$$

$$= \; \frac{2}{3(4\Delta)} \left(\frac{2\Delta^3}{24} \right) = \frac{\Delta^2}{72}$$

Thus

$$x_{mse} = 4N_5 + 4N_8 \tag{5.46}$$

$$= \frac{4\Delta^2}{72} + \frac{4\Delta^2}{144} = \frac{\Delta^2}{12} \tag{5.47}$$

Finally

$$SQNR = \frac{A^2/4}{\Delta^2/12} = \frac{(4\Delta)^2}{4} \frac{12}{\Delta^2} = 48 \tag{5.48}$$

or equivalently, $10\log_{10}(48) = 16.8$ dB.

• ⭐ SIMULATION **P15:** For the rest of the solution.

16. (a) We require $4.8 + 6k - 12 > 38$, or equivalently, $k > 7.5$. Thus we set $k = 8$ binary digits per quantization level. Recall that $r_b = f_s \log_2 Q_{level} = f_s k$ where $f_s \geq 2f_{max}$. Given that we are using an ideal anti-aliasing filter, we may take $f_{max} = 3.4$ kHz. Thus $r_b \geq (2 * 3400) * 8 = 54400$ binary digits per second.

(b) $SQNR(dB) = 10\log_{10}\left(\dfrac{3}{[\ln(1+255)]^2}\right) + 6(8) = 37.9$ dB.

17. ⭐ SIMULATION **P17:** For the full solution.

Chapter 6

Baseband Signaling

6.1 Solutions

1. ⭐ SIMULATION **P1:** For the solution. Change the input binary stream to any number of combination of 1' and 0's and observe the corresponding baseband signals.

2. ⭐ SIMULATION **P2:** For the solution. Notice that the first null-bandwidth is now $\dfrac{2}{T_b}$ i.e. twice the bandwidth requirement of polar NRZ. This is not surprising because the pulse width has been reduced by a factor of 2.

3. The minimum bandwidth requirement for a binary PCM signal

$$B_{PCM}(\min) = B \log_2 Q_{level} \qquad (6.1)$$

where B is the bandwidth of the analog signal. Since $f_s \geq 2B$,

$$B_{PCM}(\min) = \frac{f_s}{2} \log_2 Q_{level} = \frac{r_b}{2} \qquad (6.2)$$

Since

$$B_{PCM\_Multilevel} = \frac{B_{PCM\_Binary}}{n} \qquad (6.3)$$

the minimum multilevel bandwidth

$$B_{PCM\_Multilevel}(\min) = \frac{r_b}{2n} = \frac{56000}{2*4} = 7000Hz \qquad (6.4)$$

- ⭐ **SIMULATION P3:** For further details.   Experiment with the
variables to understand and appreciate their influence on the answer.

4. ⭐ **SIMULATION P4:** For the solution.   Notice that the first null
bandwidth of a 8 level polar NRZ signal is reduced by a factor of three when
compared to the first null bandwidth of a binary polar NRZ signal.   As
the range of the levels is reduced, the PSD alters accordingly.   The first
null bandwidth B_{null} will increase as T_s is reduced because $B_{null} = \frac{1}{T_s}$.   The
average signal power calculations are highlighted within the code.   Have
some fun - change the number of levels from 8 to 4 to 2.

5.

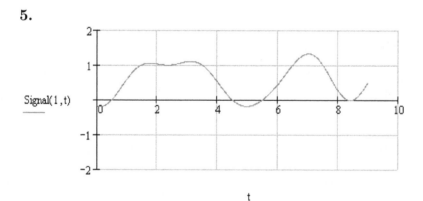

- ⭐ **SIMULATION P5:** For further details. Experiment with different
binary streams.

6. ⭐ **SIMULATION P6:** For the full solution to part(a).   Experiment
with the variables to understand and appreciate their influence on the answer.

(b) (i) The polar NRZ line code PSD is given by

$$PSD(f) = \frac{|S(f)|^2}{T_b} \tag{6.5}$$

where

$$S(f) = FT[s(t)] = T_b \frac{\sin(\pi f T_b)}{\pi f T_b} \tag{6.6}$$

Thus

$$PSD(f) = \frac{|S(f)|^2}{T_b} = T_b \left[\frac{\sin(\pi f T_b)}{\pi f T_b} \right]^2 \tag{6.7}$$

(ii) For $s(t) = \begin{cases} \cos\left(\frac{\pi t}{T_b}\right) & \text{if} \quad |t| \leq \frac{T_b}{2} \\ 0 & \text{otherwise} \end{cases}$

$$S(f) = \int_{-\infty}^{\infty} s(t)e^{-j2\pi ft}dt = \int_{-\frac{T_b}{2}}^{\frac{T_b}{2}} \cos\left(\frac{\pi t}{T_b}\right) e^{-j2\pi ft}dt \tag{6.8}$$

Using Eulers theorem

$$S(f) = \int_{-\frac{T_b}{2}}^{\frac{T_b}{2}} \cos\left(\frac{\pi t}{T_b}\right) \cos(2\pi ft)dt - j \int_{-\frac{T_b}{2}}^{\frac{T_b}{2}} \cos\left(\frac{\pi t}{T_b}\right) \sin(2\pi ft)dt \tag{6.9}$$

$$= a + jb$$

Taking the imaginary part b of this integral first by making use of the identity $\sin A \cos B = \frac{1}{2}[\sin(A - B) + \sin(A + B)]$ with $A = 2\pi ft$

$$b = \int_{-\frac{T_b}{2}}^{\frac{T_b}{2}} \cos\left(\frac{\pi t}{T_b}\right) \sin(2\pi f t) dt \tag{6.10}$$

$$= \int_{-\frac{T_b}{2}}^{\frac{T_b}{2}} \frac{1}{2}\left[\sin(2\pi f t - \frac{\pi t}{T_b}) + \sin(2\pi f t + \frac{\pi t}{T_b})\right] dt$$

$$= \int_{-\frac{T_b}{2}}^{\frac{T_b}{2}} \frac{1}{2}\left[\sin\left[2\pi\left(f - \frac{1}{2T_b}\right)t\right] + \sin\left[2\pi\left(f + \frac{1}{2T_b}\right)t\right]\right] dt \tag{6.11}$$

Now let $\alpha = \left(f - \frac{1}{2T_b}\right)$ and $\beta = \left(f + \frac{1}{2T_b}\right)$

$$b = \int_{-\frac{T_b}{2}}^{\frac{T_b}{2}} \frac{1}{2}\left[\sin\left(2\pi\alpha t\right) + \sin\left(2\pi\beta t\right)\right] dt \tag{6.12}$$

$$= \frac{1}{2}\left[-\frac{\cos\left(2\pi\alpha t\right)}{2\pi\alpha} - \frac{\cos\left(2\pi\beta t\right)}{2\pi\beta}\right]_{-\frac{T_b}{2}}^{\frac{T_b}{2}}$$

$$= \frac{1}{2}\left[-\frac{\cos\left(\pi\alpha T_b\right)}{2\pi\alpha} - \frac{\cos\left(\pi\beta T_b\right)}{2\pi\beta} - \left(-\frac{\cos\left(\pi\alpha T_b\right)}{2\pi\alpha} - \frac{\cos\left(\pi\beta T_b\right)}{2\pi\beta}\right)\right]$$

$$= \frac{1}{2}\left[-\frac{\cos\left(\pi\alpha T_b\right)}{2\pi\alpha} - \frac{\cos\left(\pi\beta T_b\right)}{2\pi\beta} + \frac{\cos\left(\pi\alpha T_b\right)}{2\pi\alpha} + \frac{\cos\left(\pi\beta T_b\right)}{2\pi\beta}\right] = 0$$

Thus

$$S(f) = \int_{-\frac{T_b}{2}}^{\frac{T_b}{2}} \cos\left(\frac{\pi t}{T_b}\right) \cos(2\pi f t) dt \tag{6.13}$$

and making use of the identity $\cos A \cos B = \frac{1}{2}\left[\cos(A-B) + \cos(A+B)\right]$

$$
S(f) = \frac{1}{2} \int_{-\frac{T_b}{2}}^{\frac{T_b}{2}} \cos(\frac{\pi t}{T_b} - 2\pi ft) + \cos(\frac{\pi t}{T_b} + 2\pi ft) dt \tag{6.14}
$$

$$
= \frac{1}{2} \int_{-\frac{T_b}{2}}^{\frac{T_b}{2}} \cos\left[2\pi\left(\frac{1}{2T_b} - f\right)t\right] + \cos\left[2\pi\left(\frac{1}{2T_b} + f\right)t\right] dt
$$

Now let $\alpha = \left(\frac{1}{2T_b} - f\right)$ and $\beta = \left(\frac{1}{2T_b} + f\right)$

$$
S(f) = \frac{1}{2} \int_{-\frac{T_b}{2}}^{\frac{T_b}{2}} \cos\left(2\pi\alpha t\right) + \cos\left(2\pi\beta t\right) dt \tag{6.15}
$$

$$
= \frac{1}{2}\left[\frac{\sin\left(2\pi\alpha t\right)}{2\pi\alpha} + \frac{\sin\left(2\pi\beta t\right)}{2\pi\beta}\right]_{-\frac{T_b}{2}}^{\frac{T_b}{2}}
$$

$$
= \frac{1}{4\pi}\left[\begin{array}{c}\left(\frac{\sin(\pi\alpha T_b)}{\alpha} + \frac{\sin(\pi\beta T_b)}{\beta}\right) \\ -\left(-\frac{\sin(\pi\alpha T_b)}{\alpha} - \frac{\sin(\pi\beta T_b)}{\beta}\right)\end{array}\right] \tag{6.16}
$$

$$
= \frac{1}{2\pi}\left(\frac{\sin\left(\pi\alpha T_b\right)}{\alpha} + \frac{\sin\left(\pi\beta T_b\right)}{\beta}\right)
$$

But $\sin\left(\pi\alpha T_b\right) = \sin\left(\frac{\pi}{2} - f\pi T_b\right) = \cos\left(f\pi T_b\right)$ and $\sin\left(\pi\beta T_b\right) = \sin\left(\frac{\pi}{2} + f\pi T_b\right) = \cos\left(f\pi T_b\right)$ so that

$$
S(f) = \frac{\cos\left(f\pi T_b\right)}{2\pi}\left[\frac{1}{\alpha} + \frac{1}{\beta}\right] = \frac{\cos\left(f\pi T_b\right)}{2\pi}\left[\frac{2T_b}{1 - f2T_b} + \frac{2T_b}{1 + f2T_b}\right] \tag{6.17}
$$

$$
= \frac{2T_b\cos\left(f\pi T_b\right)}{2\pi}\left[\frac{2}{1 - (f2T_b)^2}\right] = \frac{2T_b\cos\left(f\pi T_b\right)}{\pi}\left[\frac{1}{1 - (f2T_b)^2}\right]
$$

And finally

$$PSD(f) = \frac{|S(f)|^2}{T_b} = \frac{4T_b}{\pi^2} \left[\frac{\cos(f\pi T_b)}{1 - (f2T_b)^2} \right]^2 \tag{6.18}$$

Since $PSD(f = \frac{1}{2T_b})$ must be zero, the first null bandwidth

$$B = \frac{1}{2T_b} \tag{6.19}$$

7. ⭐ SIMULATION **P7:** For the full solution.   Experiment with the variables to understand and appreciate their influence on the answer.

8. For unipolar NRZ ,

$$P_e = Q\left(\sqrt{\frac{E_b}{N_o}}\right) = Q\left(\sqrt{10^{SNR/10}}\right) \tag{6.20}$$

where

$$SNR = 10\log_{10}\left(\frac{E_b}{N_o}\right) = 10\log_{10}\left(\frac{A^2 T_b}{N_o}\right) \tag{6.21}$$

where $N_o = 2\,(5.10^{-12}) = 1 \times 10^{-11}$ W/Hz.   Inserting the values for $A = 10^{-3}$ volts and $T_b = 10^{-4}$ s, we find $SNR = 10$ dB and hence

$$P_e = 2.7 x 10^{-4} \tag{6.22}$$

• ⭐ SIMULATION **P8:** For further details.

9. The signalling scheme is polar ZRZ for which

$$P_e = Q\left(\sqrt{\frac{E_b}{N_o}}\right) \tag{6.23}$$

To ensure $P_e \leq 10^{-6}$, we require a minimum SNR of 10.6 dB. This corresponds to $\frac{E_b}{N_o} = 11.482$. Given that $E_b = A^2 T_b$, and we know the values of N_o, A, and E_b, we find that $T_b = 0.011$ and hence the date rate

$$r_b = \frac{1}{T_b} = 87 \text{ binary digits/sec} \tag{6.24}$$

● ⭐ SIMULATION **P9:** For further details.

10. (a) For no ISI, $r = 0$ and hence the required bandwidth $B = \frac{(1+r)}{2T} = \frac{1}{2T} = \frac{1000}{2} = 500$ Hz. (b) Simply $r = (2BT - 1) = 0.8$.

⭐ SIMULATION **P10:** For further details. Experiment with the variables to understand and appreciate their influence on the answer.

11. (a) The maximum analog bandwidth $B_{\text{analog max}} = \frac{f_s}{2}$. For PCM, given that $r_b = f_s \log_2 Q_{level}$, and assuming that the overall system transfer function is raised-cosine $H_{raised}(f)$, then

$$\begin{aligned} B_{channel} &= \frac{(1+r)}{2T_b} \tag{6.25} \\ &= \frac{(1+r)}{2} r_b = \frac{(1+r) f_s \log_2 Q_{level}}{2} \end{aligned}$$

Hence

$$B_{\text{analog max}} = \frac{B_{channel}}{(1+r) \log_2 Q_{level}} \tag{6.26}$$

(b) Using a multilevel signal with M levels, the number of binary digits represented per level $n = \log_2 M$. Now $B_{channel} = \frac{(1+r)}{2T}$, where the time duration of a pulse $T = nT_b = \frac{n}{r_b}$. Thus

$$B_{channel} = \frac{(1+r)}{2nT_b} = \frac{(1+r)}{2n}r_b \tag{6.27}$$

$$= \frac{(1+r) f_s \log_2 Q_{level}}{2n}$$

so that

$$B_{analog\ max} = \frac{f_s}{2} = \frac{n B_{channel}}{(1+r) \log_2 Q_{level}} \tag{6.28}$$

i.e. we can increase the maximum analog bandwidth by a factor of $n = \log_2 M$ by switching to a multilevel signal. Notice how $B_{channel}$ is reduced by a factor of n. The penalty incurred is an increase in the probability of an error in a binary digit at the receiver.

12. The absolute bandwidth $B = \frac{(1+r)}{2T}$, where $\frac{1}{T}$ is equal to the number of pulses per second, which in this case is given by $f_s N$. Thus,

$$B = \frac{(1+r) f_s N}{2} \tag{6.29}$$

Note that we should ensure that f_s is at least twice the maximum bandwidth among the N analog signals.

13. ⭐ SIMULATION **P13:** For the full solution. Experiment with the variables to understand and appreciate their influence on the answer.

14. (a) The probability of an error in a binary digit

$$P_e = P(s_2) \int_{a_0}^{\infty} f(y|s_2)dy + P(s_1) \int_{-\infty}^{a_0} f(y|s_1)dy \tag{6.30}$$

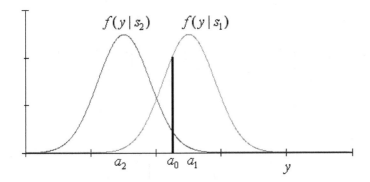

$$P_e = P(s_2)\left[1 - \int_{-\infty}^{a_0} f(y|s_2)dy\right] + P(s_1)\int_{-\infty}^{a_0} f(y|s_1)dy \qquad (6.31)$$

$$= P(s_2) - P(s_2)\int_{-\infty}^{a_0} f(y|s_2)dy + P(s_1)\int_{-\infty}^{a_0} f(y|s_1)dy$$

$$= P(s_2) + \int_{-\infty}^{a_0} \left[P(s_1)f(y|s_1) - P(s_2)f(y|s_2)\right] dy$$

To determine the value a_0 at which P_e is a minimum,

$$\frac{dP_e}{da_0} = P(s_1)f(a_0|s_1) - P(s_2)f(a_0|s_2) = 0 \qquad (6.32)$$

$$P(s_1)f(a_0|s_1) = P(s_2)f(a_0|s_2) \qquad (6.33)$$

But

$$f(y|s_1) = \frac{1}{\sigma\sqrt{2\pi}}\exp\left[-\frac{(y - a_1)^2}{2\sigma^2}\right] \qquad (6.34)$$

and

$$f(y|s_2) = \frac{1}{\sigma\sqrt{2\pi}}\exp\left[-\frac{(y - a_2)^2}{2\sigma^2}\right] \qquad (6.35)$$

Thus

$$P(s_1)\frac{1}{\sigma\sqrt{2\pi}}\exp\left[-\frac{(a_0-a_1)^2}{2\sigma^2}\right] = P(s_2)\frac{1}{\sigma\sqrt{2\pi}}\exp\left[-\frac{(a_0-a_2)^2}{2\sigma^2}\right] \quad (6.36)$$

equivalently

$$\exp\left[-\frac{(a_0-a_1)^2}{2\sigma^2}\right]\exp\left[\frac{(a_0-a_2)^2}{2\sigma^2}\right] = \frac{P(s_2)}{P(s_1)} \quad (6.37)$$

$$\exp\left[\frac{(a_0-a_2)^2}{2\sigma^2}-\frac{(a_0-a_1)^2}{2\sigma^2}\right] = \frac{P(s_2)}{P(s_1)} \quad (6.38)$$

$$\frac{(a_0-a_2)^2-(a_0-a_1)^2}{2\sigma^2} = \log_e\left(\frac{P(s_2)}{P(s_1)}\right) \quad (6.39)$$

$$= \frac{2a_0(a_1-a_2)+a_2^2-a_1^2}{2\sigma^2} \quad (6.40)$$

Thus

$$a_0 = \frac{2\sigma^2\log_e\left(\frac{P(s_2)}{P(s_1)}\right)+a_1^2-a_2^2}{2(a_1-a_2)} \quad (6.41)$$

$$= \frac{2\sigma^2}{2(a_1-a_2)}\log_e\left(\frac{P(s_2)}{P(s_1)}\right)+\frac{(a_1+a_2)(a_1-a_2)}{2(a_1-a_2)}$$

Thus

$$a_0 = \frac{\sigma^2}{(a_1-a_2)}\log_e\left(\frac{P(s_2)}{P(s_1)}\right)+\frac{(a_1+a_2)}{2} \quad (6.42)$$

$$= \frac{a_1+a_2}{2}+\frac{\sigma^2}{a_1-a_2}\ln\left(\frac{P(s_2)}{P(s_1)}\right)$$

(b) For binary antipodal signalling, with $a_1 = \sqrt{E_b}$, $a_2 = -\sqrt{E_b}$ and $\sigma^2 = \frac{N_o}{2}$

we have $a_1+a_2 = 0$ and

$$a_0 = \frac{\frac{N_o}{2}}{2\sqrt{E_b}} \log_e \left(\frac{P(s_2)}{P(s_1)} \right) \tag{6.43}$$

$$= \frac{N_o}{4\sqrt{E_b}} \log_e \left(\frac{P(s_2)}{P(s_1)} \right)$$

Note that for $P(s_2) = P(s_1)$, $a_0 = \frac{N_o}{4\sqrt{E_b}} \log_e (1) = 0$ as expected.

15. ⭐ SIMULATION P15: For the full solution. Experiment with the variables to understand and appreciate their influence on the answer.

16. (a) Signal energy $E = A^2 T_b = 4(0.5) = 2$ joules.
(b) Average output noise power is given by

$$N_{out} = \overline{n_o^2(t)} = \int_{-\infty}^{\infty} |H(f)|^2 \frac{N_o}{2} df \tag{6.44}$$

$$= \frac{N_o}{2} \int_{-\infty}^{\infty} |H(f)|^2 df$$

But

$$|H(f)|^2 = \left(\frac{1}{1 + j\frac{f}{f_o}} \right) \left(\frac{1}{1 - j\frac{f}{f_o}} \right) = \frac{1}{1 + \left(\frac{f}{f_o} \right)^2} \tag{6.45}$$

Hence

$$N_{out} = \frac{N_o}{2} \int_{-\infty}^{\infty} \frac{1}{1 + \left(\frac{f}{f_o} \right)^2} df \tag{6.46}$$

Integrating by substitution, let $u = \frac{f}{f_o}$, then $\frac{du}{df} = \frac{1}{f_o}$ and the integration limits are still $\pm\infty$.

$$
\begin{aligned}
N_{out} &= \frac{N_o}{2} \int_{-\infty}^{\infty} \left(\frac{1}{1+u^2} \right) f_o du \tag{6.47} \\
&= \frac{N_o f_o}{2} \left[\tan^{-1}(u) \right]_{-\infty}^{\infty} \\
&= \frac{N_o f_o}{2} \left(\frac{\pi}{2} - \left(\frac{\pi}{2} \right) \right) \\
&= \frac{N_o f_o \pi}{2} = \frac{N_o \pi}{2} \left(\frac{1}{2\pi RC} \right) \\
&= \frac{N_o}{4RC} = \frac{N_o}{4t_c} = \frac{10^{-11}}{4(0.3)} = 8.33 x 10^{-12} \text{ W}
\end{aligned}
$$

The output of the RC filter at time t is given by

$$
s(t) = A \left(1 - e^{-\frac{t}{t_c}} \right) \tag{6.48}
$$

Hence the output signal power at time T is given by

$$
\begin{aligned}
P_{out} &= s(T)^2 = \left[A \left(1 - e^{-\frac{T}{t_c}} \right) \right]^2 \tag{6.49} \\
&= \left[2 \left(1 - e^{-\frac{0.5}{0.3}} \right) \right]^2 = 2.632 \text{ W}
\end{aligned}
$$

Hence,

$$
\left(\frac{P_{out}}{N_{out}} \right)_{RCFilter} = \frac{2.632}{8.33 x 10^{-12}} = 3.16 x 10^{11} \simeq 115 \text{ dB} \tag{6.50}
$$

(c) Using a matched filter,

$$
\frac{P_{out}}{N_{out}} = \frac{2E}{N_o} = \frac{2(2)}{10^{-11}} = 4 x 10^{11} \simeq 116 \text{ dB} \tag{6.51}
$$

• ⭐ SIMULATION **P16:** For part (d).

17. Under a noiseless channel,

$$a_1 = \int_0^T s_1(t)dt = \int_0^T Adt = AT \tag{6.52}$$

$$a_2 = \int_0^T s_2(t)dt = \int_0^T -Adt = -AT \tag{6.53}$$

Following the standard derivation

$$P_e = Q\left(\frac{a_1 - a_2}{2\sigma}\right) \tag{6.54}$$

$$= Q\left(\frac{2AT}{2} * \sqrt{\frac{2}{N_oT}}\right)$$

$$= Q\left(\sqrt{\frac{2A^2T}{N_o}}\right) = Q\left(\sqrt{\frac{2E_b}{N_o}}\right)$$

where the energy per binary digit $E_b = A^2T$.

18. The average or expected value of a random process $X(t)$ is defined by

$$E[X(t)] = \int_{-\infty}^{\infty} x f_X(x)dx \tag{6.55}$$

Note that $X(t)$ is a random variable for a fixed value of t. The autocorrelation of $X(t)$ is defined by

$$R_X(t_1, t_2) = E[X(t_1)X(t_2)] \tag{6.56}$$

For a wide-sense (or weakly) stationary random process

$$R_X(\tau) = E[X(t)X(t + \tau)] \tag{6.57}$$

where $\tau = t_2 - t_1$. From the Wiener-Khinchine theorem, the PSD of a wide-sense stationary random process is the FT of its autocorrelation function. Taking the inverse FT of the white noise PSD $\frac{N_o}{2}$, we get

$$R_X(\tau) = \frac{N_o}{2}\delta(\tau) \qquad (6.58)$$

as shown below.

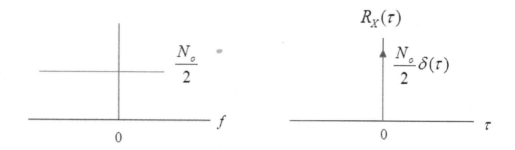

Consider now the integrate and dump receiver: The output

$$n_o = \int_0^{T_b} n(t)dt \qquad (6.59)$$

where n_o is a gaussian random variable.

$$
\begin{aligned}
E\left[n_o\right] &= E\left[\int_0^{T_b} n(t)dt\right] \qquad (6.60)\\
&= \int_0^{T_b} E\left[n(t)\right] dt
\end{aligned}
$$

but $E\left[n(t)\right] = 0$ (zero mean) and thus $E\left[n_o\right] = 0$. Recall the variance

$$\sigma^2 = E\left[X^2\right] - E\left[X\right]^2$$

Thus, the variance

$$\sigma^2 = E\left[n_o^2\right] - E\left[n_o\right]^2 = E\left[n_o^2\right] \tag{6.61}$$

$$= E\left[\left[\int_0^{T_b} n(t)dt\right]^2\right] = \int_0^{T_b}\int_0^{T_b} E\left[n(t)n(\tau)\right] dtd\tau$$

But

$$E\left[n(t)n(\tau)\right] = R_X(t-\tau) = \frac{N_o}{2}\delta(t-\tau) \tag{6.62}$$

Thus

$$\sigma^2 = \int_0^{T_b}\int_0^{T_b} \frac{N_o}{2}\delta(t-\tau)dtd\tau \tag{6.63}$$

But

$$\int_0^{T_b} \frac{N_o}{2}\delta(t-\tau)dt \tag{6.64}$$

$$= \frac{N_o}{2}\int_0^{T_b} \delta(t-\tau)dt = \frac{N_o}{2}$$

since $\delta(t-\tau) = \delta(0) = 1$ only when $t = \tau$. Hence

$$\sigma^2 = \int_0^{T_b} \frac{N_o}{2}d\tau = \frac{N_o}{2}\int_0^{T_b} d\tau = \frac{N_o T_b}{2} \tag{6.65}$$

or equivalently

$$\sigma = \sqrt{\frac{N_o T_b}{2}} \tag{6.66}$$

(b) **Hint:** Simply modify the simulation code in ⭐ SIMULATION **SignalCorrelator**

⭐ SIMULATION **P18:** For the solution.

19. (a) The noise component output n_o of the correlator at time $t = T_b$ is given by

$$n_o = \int_0^{T_b} n(t)s(t)dt \tag{6.67}$$

The variance

$$\sigma^2 = E\left[n_o^2\right] = E\left[\left[\int_0^{T_b} n(t)s(t)dt\right]^2\right] \tag{6.68}$$

$$= \int_0^{T_b}\int_0^{T_b} E\left[n(t)n(\tau)\right]s(t)s(\tau)dtd\tau$$

But

$$E\left[n(t)n(\tau)\right] = R_X(t-\tau) = \frac{N_o}{2}\delta(t-\tau) \tag{6.69}$$

Therefore

$$\sigma^2 = \int_0^{T_b}\int_0^{T_b} \frac{N_o}{2}\delta(t-\tau)s(t)s(\tau)dtd\tau \tag{6.70}$$

But

$$\int_0^{T_b} \frac{N_o}{2}\delta(t-\tau)dt = \frac{N_o}{2} \tag{6.71}$$

Hence $\sigma^2 = \frac{N_o}{2}\int_0^{T_b} s^2(\tau)d\tau = \frac{N_o E_b}{2}$.

(b) Using the same analysis as in part(a) with $s(t)$ replaced by $\phi(t)$, we find that

$$\sigma^2 = \frac{N_o}{2}\int_0^{T_b} \phi^2(\tau)d\tau = \frac{N_o}{2} \tag{6.72}$$

20. ⭐ SIMULATION **P20:** For parts (a), (b), (c).

(i) Consider first scenario where $s_1(t) = a(t)$ and $s_2(t) = b(t)$. If $r(t) = s_1(t)$, then we find $y_1 = +E_b$ and $y_2 = 0$. If $r(t) = s_2(t)$, then we find $y_1 = 0$ and $y_2 = +E_b$ (refer to ⭐ SIMULATION **P20** as required).

With noise,

$$r(t) = s_i(t) + n(t) \tag{6.73}$$

where $i = 1, 2$. If $s_1(t)$ is transmitted, then $y_1 = E_b + n_1$ and $y_2 = n_2$, where the noise components

$$n_1 = \int_0^{T_b} n(t) s_1(t) dt \tag{6.74}$$

and

$$n_2 = \int_0^{T_b} n(t) s_2(t) dt \tag{6.75}$$

From the result of the previous question, the variance of n_1 and n_2 is $\sigma^2 = \frac{N_o}{2} E_b$. By symmetry,

$$\begin{aligned} P_e &= P(y_2 > y_1) = P(n_2 > E_b + n_1) \\ &= P(n_2 - n_1 > E_b) = P(n > E_b) \end{aligned} \tag{6.76}$$

where

$$n = n_2 - n_1 \tag{6.77}$$

The mean of the random variable n is given by

$$\begin{aligned} E\left[(n_2 - n_1)\right] &\\ &= E\left[\int_0^{T_b} n(t) s_2(t) dt - \int_0^{T_b} n(\tau) s_1(\tau) d\tau\right] \\ &= \int_0^{T_b} E\left[n(t)\right] s_2(t) dt - \int_0^{T_b} E\left[n(\tau)\right] s_1(\tau) d\tau \end{aligned} \tag{6.78}$$

but $E\left[n(t)\right] = E\left[n(\tau)\right] = 0$ and thus $E\left[(n_2 - n_1)\right] = 0$. This not surprising because n_1 and n_2 are zero-mean gaussian random variables. Its variance

$$
\begin{aligned}
\sigma_n^2 &= E\left[(n_2 - n_1)^2\right] \\
&= E\left[n_2^2\right] + E\left[n_1^2\right] - 2E\left[n_2 n_1\right] \\
&= \frac{N_o}{2}E_b + \frac{N_o}{2}E_b - 2E\left[n_2 n_1\right]
\end{aligned}
\tag{6.79}
$$

However

$$
E\left[n_2 n_1\right] = \int_0^{T_b} \int_0^{T_b} E\left[n(t)n(\tau)\right] s_2(t)s_1(\tau)dtd\tau
\tag{6.80}
$$

But

$$
E\left[n(t)n(\tau)\right] = \frac{N_o}{2}\delta(t - \tau)
\tag{6.81}
$$

Hence

$$
\begin{aligned}
E\left[n_2 n_1\right] &= \frac{N_o}{2} \int_0^{T_b} \int_0^{T_b} \delta(t - \tau)s_2(t)s_1(\tau)dtd\tau \\
&= \frac{N_o}{2} \int_0^{T_b} s_2(\tau)s_1(\tau)d\tau
\end{aligned}
\tag{6.82}
$$

Recall from part(a), $\int_0^{T_b} s_2(\tau)s_1(\tau)d\tau = 0$ (i.e. the signals are orthogonal) and therefore

$$
E\left[n_2 n_1\right] = 0
\tag{6.83}
$$

Thus

$$
\sigma_n^2 = \frac{N_o}{2}E_b + \frac{N_o}{2}E_b = N_o E_b
\tag{6.84}
$$

Given that n is a zero-mean gaussian random variable with variance $N_o E_b$, the

$$P_e = P\left(n > E_b\right) = \int\limits_{E_b}^{\infty} \frac{1}{\sigma_n\sqrt{2\pi}} e^{\frac{-n^2}{2\sigma_n^2}} dn \tag{6.85}$$

Let $\frac{-u^2}{2} = \frac{-n^2}{2\sigma_n^2}$ so that $u = \frac{n}{\sigma_n}$ and $\frac{du}{dn} = \frac{1}{\sigma_n}$. The integration limits are changed to $\frac{E_b}{\sigma_n}$ and ∞ so that

$$P_e = \int\limits_{\frac{E_b}{\sigma_n}}^{\infty} \frac{1}{\sqrt{2\pi}} e^{\frac{-u^2}{2}} du = Q\left(\frac{E_b}{\sigma_n}\right) \tag{6.86}$$

$$= Q\left(\frac{E_b}{\sqrt{N_o E_b}}\right) = Q\left(\sqrt{\frac{E_b}{N_o}}\right)$$

(ii) Now consider second scenario where $s_1(t) = a(t)$ and $s_2(t) = -a(t)$. If $r(t) = s_1(t)$, then $y_1 = +E_b$ and $y_2 = -E_b$. If $r(t) = s_2(t)$, then $y_1 = -E_b$ and $y_2 = +E_b$ (refer to ⭐ SIMULATION **P20** as required). If $s_1(t)$ is transmitted, then $y_1 = E_b + n_1$ and $y_2 = -E_b + n_2$. Thus

$$\begin{aligned} P_e &= P\left(y_2 > y_1\right) = P\left(-E_b + n_2 > E_b + n_1\right) \tag{6.87}\\ &= P\left(n_2 - n_1 > 2E_b\right) = P\left(n > 2E_b\right) \end{aligned}$$

where

$$n = n_2 - n_1 \tag{6.88}$$

Given that n_1 and n_2 are zero-mean gaussian random variables, the mean of the random variable n is zero. Its variance

$$\begin{aligned} \sigma_n^2 &= E\left[(n_2 - n_1)^2\right] \tag{6.89}\\ &= E\left[n_2^2\right] + E\left[n_1^2\right] - 2E\left[n_2 n_1\right]\\ &= \frac{N_o}{2} E_b + \frac{N_o}{2} E_b - 2E\left[n_2 n_1\right] \end{aligned}$$

However

$$E\left[n_2 n_1\right] = \int_0^{T_b} \int_0^{T_b} E\left[n(t)n(\tau)\right] s_2(t)s_1(\tau)dt d\tau \tag{6.90}$$

But

$$E\left[n(t)n(\tau)\right] = \frac{N_o}{2}\delta(t-\tau) \tag{6.91}$$

Hence

$$\begin{aligned} E\left[n_2 n_1\right] &= \frac{N_o}{2} \int_0^{T_b} \int_0^{T_b} \delta(t-\tau)s_2(t)s_1(\tau)dt d\tau \\ &= \frac{N_o}{2} \int_0^{T_b} s_2(\tau)s_1(\tau)d\tau \end{aligned} \tag{6.92}$$

Recall from part(b), $\int_0^{T_b} s_2(\tau)s_1(\tau)d\tau = -E_b$ and therefore

$$E\left[n_2 n_1\right] = -\frac{N_o E_b}{2} \tag{6.93}$$

Thus

$$\sigma_n^2 = \frac{N_o}{2}E_b + \frac{N_o}{2}E_b - 2\left(-\frac{N_o E_b}{2}\right) = 2N_o E_b \tag{6.94}$$

Given that n is a zero-mean gaussian random variable with variance $2N_o E_b$, the

$$P_e = P\left(n > 2E_b\right) = \int_{2E_b}^{\infty} \frac{1}{\sigma_n\sqrt{2\pi}} e^{\frac{-n^2}{2\sigma_n^2}} dn \tag{6.95}$$

Let $\frac{-u^2}{2} = \frac{-n^2}{2\sigma_n^2}$ so that $u = \frac{n}{\sigma_n}$ and $\frac{du}{dn} = \frac{1}{\sigma_n}$. The integration limits are changed to $\frac{2E_b}{\sigma_n}$ and ∞ so that

$$P_e = \int_{\frac{2E_b}{\sigma_n}}^{\infty} \frac{1}{\sqrt{2\pi}} e^{\frac{-u^2}{2}} du = Q\left(\frac{2E_b}{\sigma_n}\right) \tag{6.96}$$

$$= Q\left(\frac{2E_b}{\sqrt{2N_oE_b}}\right) = Q\left(\sqrt{\frac{2E_b}{N_o}}\right)$$

(iii) By symmetry, if $s_1(t) = b(t)$ and $s_2(t) = -b(t)$, then $P_e = Q\left(\sqrt{\frac{2E_b}{N_o}}\right)$ as expected for antipodal signals.

21. Recall that the bandwidth of a PCM signal is given by $B_{PCM} \geq \frac{r_b}{2}$. However, referring to the polar NRZ line code PSD, the null bandwidth $B_{null} = r_b$. Clearly, polar NRZ signalling will require twice the minimum possible bandwidth (that could be achieved in theory using $\frac{\sin(x)}{x}$ shaped pulses).

⭐ SIMULATION **P21:** For the rest of the solution.

Chapter 7

Bandpass Signaling

7.1 Solutions

1. ▶ VIDEO SOLUTION **P1:** The probability of an error in a binary digit for BPSK is given by

$$P_e = Q\left(\sqrt{\frac{2E_b}{N_o}}\right) \tag{7.1}$$

where

$$\left(\frac{E_b}{N_o}\right) dB = 10\log_{10}\left(\frac{E_b}{N_o}\right) = 0 \text{ dB} \tag{7.2}$$

Thus $\left(\dfrac{E_b}{N_o}\right) = 10^{0/10} = 1$ and so we find

$$P_e = Q\left(\sqrt{2}\right) = Q\left(1.41\right) = 0.0793 \tag{7.3}$$

Thus the BSC crossover probability $\alpha = 0.0793$. Hence probability of correctly receiving the sequence $1001 = (1-\alpha)^4 = 0.72$.

• ★ SIMULATION **P1:** For further details.

- Experiment with the variables to understand and appreciate their influence on the answer.

2. ▶ VIDEO SOLUTION P2: (a) A BPSK signal in general is given by $s_1(t) = A \cos(2\pi f_c t)$. Thus by inspection $A = 1$ Volts and $f_c = 10$ Hz.

(b) $E_b = \dfrac{A^2 T_b}{2} = 0.25$ Joules.

(c) Coordinates $= \pm \sqrt{E_b} = \pm 0.5$.

3. ▶ VIDEO SOLUTION P3: $r_c = 90000$ binary digits/sec, $A = 0.005$ vols, $N_0 = 10^{-11}$ W/Hz

The time duration of a MPSK signal

$$T_s = T_b R_{code} \log_2(M) = T_c \log_2(M) = \frac{1}{r_c} \log_2(M) \qquad (7.4)$$

which for a BPSK signal, $T_s = \frac{1}{r_c}$ because $M = 2$. The amplitude of a MPSK signal

$$A = \sqrt{\frac{2E_s}{T_s}} = \sqrt{\frac{2E_b R_{code} \log_2(M)}{T_b R_{code} \log_2(M)}} = \sqrt{\frac{2E_b}{T_b}} \qquad (7.5)$$

Given that there is no further information about the communication system, we have to assume that error-control was not being used ($R_{code} = 1$) so that

$$T_s = T_b R_{code} \log_2(M) = T_b(1) \log_2(2) = T_b \qquad (7.6)$$

Thus

$$E_b = \frac{A^2 T_b}{2} = \frac{A^2}{2r_c} \qquad (7.7)$$

Hence

$$\frac{E_b}{N_o} = \frac{A^2}{2r_c N_o} \tag{7.8}$$

Thus we find

$$P_e = Q\left(\sqrt{\frac{2E_b}{N_o}}\right) = Q\left(\sqrt{\frac{A^2}{r_c N_o}}\right) = Q(5.27) = 6.8\text{x}10^{-8} \tag{7.9}$$

(b) Number of seconds in one hour $= 60 * 60 = 3600$
Average number of errors per hour $= 90000 * 6.8 \text{ x } 10^{-8} * 3600 = 22$.

(c) Average signal power $= E_b r_c = 1.25 \text{ x } 10^{-5}$ watts.

• ⭐ SIMULATION P3: For further details.   Experiment with the variables to understand and appreciate their influence on the answer.

4. ▶ VIDEO SOLUTION P4: (a) $P_e = Q\left(\sqrt{\frac{2E_b}{N_o}}\right)$, where $\left(\frac{E_b}{N_o}\right) dB =$
$10\log_{10}\left(\frac{E_b}{N_o}\right) = 4$ dB. Thus $\left(\frac{E_b}{N_o}\right) = 10^{4/10} = 2.51$, from which we find
$P_e = Q\left(\sqrt{2(2.51)}\right) = Q(2.24) = 0.0125$.

(b) $P_e = Q\left(\sqrt{\frac{2E_b R_{code}}{N_o}}\right) = Q\left(\sqrt{2(2.51)\,0.5}\right) = Q(1.58) = 0.0571$.

• ⭐ SIMULATION P4: For further details.   Experiment with the variables to understand and appreciate their influence on the answer.

5. ▶ VIDEO SOLUTION P5:

$$x_h = \int_0^{T_b} s_1(t)\phi_1(t)dt \tag{7.10}$$

Recall BPSK is given by $s_1(t) = \sqrt{\dfrac{2E_b}{T_b}} \cos(2\pi f_c t)$, thus

$$x_h = \frac{2\sqrt{E_b}}{T_b} \int_0^{T_b} \cos(2\pi f_c t) \cos(2\pi f_c t + \theta)\, dt \tag{7.11}$$

but $\cos(2\pi f_c t + \theta) = \cos(2\pi f_c t)\cos(\theta) - \sin(2\pi f_c t)\sin(\theta)$, therefore

$$x_h = \frac{2\sqrt{E_b}}{T_b} \int_0^{T_b} \cos(2\pi f_c t)\left[\cos(2\pi f_c t)\cos(\theta) - \sin(2\pi f_c t)\sin(\theta)\right] dt \tag{7.12}$$

$$= \frac{2\sqrt{E_b}}{T_b} \int_0^{T_b} \left[\cos^2(2\pi f_c t)\cos(\theta) - \cos(2\pi f_c t)\sin(2\pi f_c t)\sin(\theta)\right] dt$$

but $\cos^2(2\pi f_c t) = \dfrac{1 + \cos(4\pi f_c t)}{2}$, $\sin(4\pi f_c t) = 2\cos(2\pi f_c t)\sin(2\pi f_c t)$, thus

$$x_h = \frac{2\sqrt{E_b}}{T_b} \int_0^{T_b} \left[\frac{\cos\theta}{2} + \frac{\cos\theta\cos(4\pi f_c t)}{2} - \frac{\sin(4\pi f_c t)}{2}\sin\theta\right] dt \tag{7.13}$$

$$= \frac{\sqrt{E_b}}{T_b} \int_0^{T_b} \left[\cos\theta + \cos\theta\cos(4\pi f_c t) - \sin(4\pi f_c t)\sin\theta\right] dt$$

Now the integral is over time t and $\cos\theta$ is constant within this integeral, so that

$$\int_0^{T_b} \cos\theta\, dt = \cos\theta \int_0^{T_b} dt = t\cos\theta \tag{7.14}$$

$$\int_0^{T_b} \cos\theta \cos(4\pi f_c t)\, dt = \cos\theta \int_0^{T_b} \cos(4\pi f_c t)\, dt = \cos\theta \frac{\sin(4\pi f_c t)}{4\pi f_c} \quad (7.15)$$

$$\int_0^{T_b} -\sin(4\pi f_c t)\sin\theta\, dt = -\sin\theta \int_0^{T_b} \sin(4\pi f_c t)\, dt = \sin\theta \frac{\cos(4\pi f_c t)}{4\pi f_c} \quad (7.16)$$

Thus

$$x_h = \frac{\sqrt{E_b}}{T_b}\left[t\cos\theta + \frac{\cos\theta\sin(4\pi f_c t)}{4\pi f_c} + \frac{\cos(4\pi f_c t)}{4\pi f_c}\sin\theta\right]_0^{T_b} \quad (7.17)$$

Recall that the time duration of a BPSK signal $T_b = n_c T_c = \frac{n_c}{f_c}$ where n_c is an integer. Thus $\sin(4\pi f_c T_b) = \sin\left(4\pi f_c \frac{n_c}{f_c}\right) = \sin(4\pi n_c) = 0$ and similarly $\cos(4\pi f_c T_b) = \cos(4\pi n_c) = 1$. Thus

$$x_h = \frac{\sqrt{E_b}}{T_b}\left[T_b\cos\theta + \frac{\sin\theta}{4\pi f_c} - \frac{\sin\theta}{4\pi f_c}\right] = \sqrt{E_b}\cos\theta \quad (7.18)$$

For $\theta = 0$, $x_h = \sqrt{E_b}$ as expected. Similarly by symmetry, for $s_2(t) = -\sqrt{\frac{2E_b}{T_b}}\cos(2\pi f_c t)$, we have $x_h = -\sqrt{E_b}\cos\theta$.

Following the derivation in the text and replacing $\sqrt{E_b}$ by $\sqrt{E_b}\cos\theta$, we find that $P_e = Q\left(\sqrt{\frac{2E_b}{N_o}}\cos\theta\right)$. For $\theta = 0$, $P_e = Q\left(\sqrt{\frac{2E_b}{N_o}}\right)$ as expected.

(b) For $Q\left(\sqrt{\frac{2E_b}{N_o}}\cos\theta\right) = \frac{1}{2}$, require $\sqrt{\frac{2E_b}{N_o}}\cos\theta = 0$ which is satisfied for $\theta = 90$ degrees.

(c) $Q\left(\sqrt{\frac{2E_b}{N_o}}\cos 0\right) = 0.0002$ and $Q\left(\sqrt{\frac{2E_b}{N_o}}\cos\theta\right) = 0.0012$, where $\sqrt{\frac{2E_b}{N_o}} = 3.49$, $\sqrt{\frac{2E_b}{N_o}}\cos\theta = 3.03$, $\cos\theta = \frac{3.03}{3.49}$. Therefore $\theta = 29.8$ degrees.

- ⭐ SIMULATION **P5:** For further details.   Experiment with the variables to understand and appreciate their influence on the answer.

6. ▶ VIDEO SOLUTION **P6:** For $s_1(t) = \sqrt{\dfrac{2E_b}{T_b}} \cos{(2\pi f_c t)}$, recall that $x_h = \displaystyle\int_0^{T_b} s_1(t)\phi_1(t)dt = \sqrt{E_b}$.

For $s_2(t) = -\sqrt{\dfrac{E_b}{2T_b}} \cos{(2\pi f_c t)}$, we find that

$$x_h = \int_0^{T_b} s_2(t)\phi_1(t)dt = \int_0^{T_b} -\sqrt{\frac{E_b}{2T_b}} \cos{(2\pi f_c t)} \sqrt{\frac{2}{T_b}} \cos{(2\pi f_c t)}\, dt \quad (7.19)$$

$$= -\sqrt{\frac{2}{T_b}}\sqrt{\frac{E_b}{2T_b}}\int_0^{T_b} \cos^2{(2\pi f_c t)}\, dt = -\frac{\sqrt{E_b}}{T_b}\int_0^{T_b} \cos^2{(2\pi f_c t)}\, dt$$

but $\cos^2{(2\pi f_c t)} = \dfrac{1 + \cos{(4\pi f_c t)}}{2}$, so that

$$x_h = -\frac{\sqrt{E_b}}{T_b}\int_0^{T_b} \frac{1}{2} + \frac{\cos{(4\pi f_c t)}}{2}\, dt \qquad (7.20)$$

$$= -\frac{\sqrt{E_b}}{2T_b}\left[t + \frac{\sin{(4\pi f_c t)}}{4\pi f_c}\right]_0^{T_b} = -\frac{\sqrt{E_b}}{2T_b}\left[T_b + \frac{\sin{(4\pi f_c T_b)}}{4\pi f_c}\right]$$

But $\sin{(4\pi f_c T_b)} = \sin{\left(4\pi f_c \dfrac{n}{f_c}\right)} = \sin{(4\pi n)} = 0$, where n is an integer. Thus

$$x_h = -\frac{\sqrt{E_b}}{2} \qquad (7.21)$$

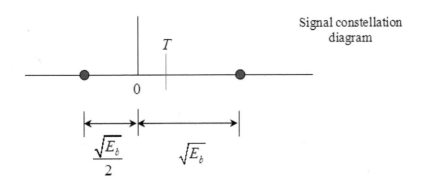

Signal constellation diagram

Optimum threshold is half way between the two signal points because the PDF of each point is the same (i.e. noise effects each signal point in the same way).

Hence $T = \dfrac{-\dfrac{\sqrt{E_b}}{2} + \sqrt{E_b}}{2} = \dfrac{\sqrt{E_b}}{4}$.

7. VIDEO SOLUTION P7:

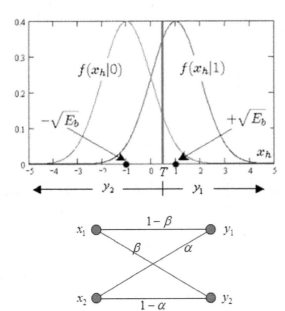

$$\alpha = P(y_1|x_2) = \int_T^\infty f(x_h|0)dx_h \tag{7.22}$$

$$= \int_T^\infty \frac{1}{\sigma\sqrt{2\pi}}\exp\left[-\frac{\left(x_h + \sqrt{E_b}\right)^2}{2\sigma^2}\right]dx_h$$

Following the derivation in the text, we find

$$\alpha = \int_{\frac{T+\sqrt{E_b}}{\sigma}}^\infty \frac{1}{\sqrt{2\pi}}\exp\left[-\frac{z^2}{2}\right]dz = Q\left(\frac{T+\sqrt{E_b}}{\sigma}\right) \tag{7.23}$$

Similarly,

$$\beta = P(y_2|x_1) = \int_{-\infty}^T f(x_h|1)dx_h \tag{7.24}$$

$$= \int_{-\infty}^T \frac{1}{\sigma\sqrt{2\pi}}\exp\left[-\frac{\left(x_h - \sqrt{E_b}\right)^2}{2\sigma^2}\right]dx_h$$

$$\beta = \int_{-\infty}^{\frac{T-\sqrt{E_b}}{\sigma}} \frac{1}{\sqrt{2\pi}}\exp\left[-\frac{z^2}{2}\right]dz = Q\left(\frac{\sqrt{E_b}-T}{\sigma}\right) \tag{7.25}$$

Thus, the probability of error

$$P_e = P(y_1|x_2)P(x_2) + P(y_2|x_1)P(x_1) \tag{7.26}$$

Assuming $P(x_1) = P(x_2) = \dfrac{1}{2}$

$$P_e = \frac{1}{2}\left[Q\left(\frac{T+\sqrt{E_b}}{\sigma}\right) + Q\left(\frac{\sqrt{E_b}-T}{\sigma}\right)\right] \tag{7.27}$$

where $\sigma = \sqrt{\frac{N_o}{2}}$. As a check, we find that for $T = 0$,

$$
\begin{aligned}
P_e &= \frac{1}{2}\left[Q\left(\frac{\sqrt{E_b}}{\sigma}\right) + Q\left(\frac{\sqrt{E_b}}{\sigma}\right)\right] \\
&= Q\left(\frac{\sqrt{E_b}}{\sigma}\right) = Q\left(\sqrt{\frac{2E_b}{N_o}}\right)
\end{aligned}
\tag{7.28}
$$

8. 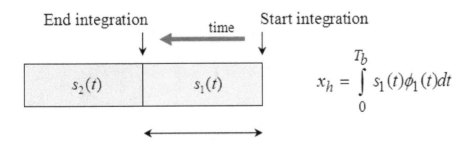VIDEO SOLUTION **P8:** With Gray coding, $P_e = \dfrac{P_s}{\log_2 M}$ e.g. for 8-PSK, $P_e = \dfrac{P_s}{\log_2 8} = \dfrac{P_s}{3}$.

9. VIDEO SOLUTION **P9:** (a) Signals arrive in the order $s_1(t)$ first and then $s_2(t)$. If there is no timing error at the integrator, the diagram below illustrates the integration process.

End integration   time   Start integration

| $s_2(t)$ | $s_1(t)$ |

$$x_h = \int_0^{T_b} s_1(t)\phi_1(t)dt$$

However, if the integrator is late by (kT_b) seconds, then the integration process is illustrated below.

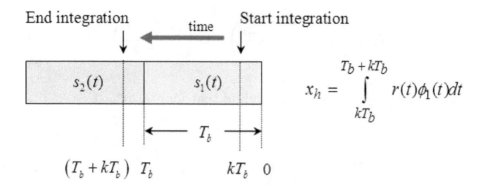

$$x_h = \int_{kT_b}^{T_b + kT_b} r(t)\phi_1(t)dt$$

Hence in this case, the integration occurs over a part of $s_1(t)$ and $s_2(t)$ as follows.

$$x_h = \int_{kT_b}^{T_b} s_1(t)\phi_1(t)dt + \int_{T_b}^{T_b+kT_b} s_2(t)\phi_1(t)dt \qquad (7.29)$$

$$= \frac{\sqrt{E_b}}{T_b}\left[t + \frac{\sin\left(4\pi f_c t\right)}{4\pi f_c}\right]_{kT_b}^{T_b}$$

$$+ \frac{-\sqrt{E_b}}{T_b}\left[t + \frac{\sin\left(4\pi f_c t\right)}{4\pi f_c}\right]_{T_b}^{T_b+kT_b}$$

But

$$\sin\left(4\pi f_c T_b\right) = \sin\left(4\pi f_c \frac{n}{f_c}\right) = \sin\left(4\pi n\right) = 0 \qquad (7.30)$$

where n is an integer and $\cos\left(4\pi n\right) = 1$.

$$x_h = \frac{\sqrt{E_b}}{T_b}(T_b - kT_b) - \frac{\sqrt{E_b}}{T_b}(T_b + kT_b - T_b) \qquad (7.31)$$

$$= \sqrt{E_b}(1 - k) - \sqrt{E_b}(k) = \sqrt{E_b}(1 - k - k) = \sqrt{E_b}(1 - 2k)$$

(b) For signal $s_2(t)$ is followed by $s_1(t)$, by symmetry we find

$$x_h = -\frac{\sqrt{E_b}}{T_b}(T_b - kT_b) + \frac{\sqrt{E_b}}{T_b}(T_b + kT_b - T_b) \tag{7.32}$$

$$= -\sqrt{E_b}(1-k) + \sqrt{E_b}(k) = -\sqrt{E_b}(1-k-k) = -\sqrt{E_b}(1-2k)$$

(c) For signal $s_1(t)$ is followed by $s_1(t)$, by symmetry we find

$$x_h = \frac{\sqrt{E_b}}{T_b}(T_b - kT_b) + \frac{\sqrt{E_b}}{T_b}(T_b + kT_b - T_b) \tag{7.33}$$

$$= \sqrt{E_b}(1-k) + \sqrt{E_b}(k) = \sqrt{E_b}(1-k+k) = \sqrt{E_b}$$

(d) For signal $s_2(t)$ is followed by $s_2(t)$, by symmetry we find

$$x_h = -\frac{\sqrt{E_b}}{T_b}(T_b - kT_b) - \frac{\sqrt{E_b}}{T_b}(T_b + kT_b - T_b) \tag{7.34}$$

$$= -\sqrt{E_b}(1-k) - \sqrt{E_b}(k) = -\sqrt{E_b}(1-k+k) = -\sqrt{E_b} \tag{7.35}$$

(e) Assuming that each signal is equally likely to be received, refer to the table below.

Pair received	x_h	Probability
$s_1(t)$ is followed by $s_2(t)$	$\sqrt{E_b}(1-2k)$	1/4
$s_2(t)$ is followed by $s_1(t)$	$-\sqrt{E_b}(1-2k)$	1/4
$s_1(t)$ is followed by $s_1(t)$	$\sqrt{E_b}$	1/4
$s_2(t)$ is followed by $s_2(t)$	$-\sqrt{E_b}$	1/4

We know that when $x_h = \pm\sqrt{E_b}$, the probability of an error in a binary digit is $Q\left(\sqrt{\frac{2E_b}{N_o}}\right)$. By symmetry, if $x_h = \pm\sqrt{E_b}(1-2k)$, then the probability of an error in a binary digit $= Q\left((1-2k)\sqrt{\frac{2E_b}{N_o}}\right)$. Thus the overall probability of an error in a binary digit

$$P_e = \frac{1}{2}Q\left(\sqrt{\frac{2E_b}{N_o}}\right) + \frac{1}{2}Q\left((1-2k)\sqrt{\frac{2E_b}{N_o}}\right) \tag{7.36}$$

10. ▶VIDEO SOLUTION **P10:** For $s_1(t)$, the input to the low-pass filter is given by

$$
\begin{aligned}
s_1(t)\phi_1(t) &= k\cos\left(2\pi f_c t\right)\cos\left(2\pi f_c t + \theta\right) && (7.37)\\
&= k\cos\left(2\pi f_c t\right)\left[\cos\left(2\pi f_c t\right)\cos\theta - \sin\left(2\pi f_c t\right)\sin\theta\right]\\
&= k\cos^2\left(2\pi f_c t\right)\cos\theta - k\cos\left(2\pi f_c t\right)\sin\left(2\pi f_c t\right)\sin\theta\\
&= k\cos^2\left(2\pi f_c t\right)\cos\theta - \frac{k\sin\left(4\pi f_c t\right)\sin\theta}{2}
\end{aligned}
$$

but

$$
\cos^2\left(2\pi f_c t\right) = \frac{1}{2} + \frac{\cos\left(4\pi f_c t\right)}{2} \qquad (7.38)
$$

Thus

$$
s_1(t)\phi_1(t) = \frac{k\cos\theta}{2} + \frac{k\cos\left(4\pi f_c t\right)\cos\theta}{2} - \frac{k\sin\left(4\pi f_c t\right)\sin\theta}{2} \qquad (7.39)
$$

Using a low-pass filter, the high frequency components are eliminated. Thus the output of the low-pass filter is $\frac{k\cos\theta}{2}$. Similarly for the input signal $s_2(t)$, the output will be $\frac{-k\cos\theta}{2}$. Three values of θ of interest are listed below.

Condition	Input Signal $s_1(t)$	Input Signal $s_2(t)$	Comment
$\theta = 0$	$\dfrac{k}{2}$	$-\dfrac{k}{2}$	Ideal output
$\theta = 180$	$-\dfrac{k}{2}$	$\dfrac{k}{2}$	Binary digits inverted
$\theta = 90$	0	0	Unable to decode

11. ▶VIDEO SOLUTION **P11:** (a)

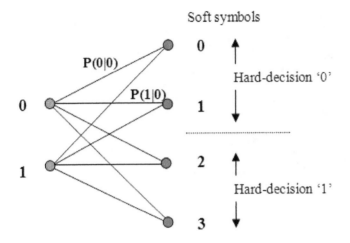

Soft symbols

(b)

$$P(0|0) = \int_{-\infty}^{-a} f(x_h|0)\, dx_h \qquad (7.40)$$

where

$$f(x_h|0) = \frac{1}{\sigma\sqrt{2\pi}} \exp\left[-\frac{\left(x_h + \sqrt{E_b}\right)^2}{2\sigma^2}\right] \qquad (7.41)$$

let $z = \dfrac{\left(x_h + \sqrt{E_b}\right)}{\sigma}$, so that $\dfrac{dz}{dx_h} = \dfrac{1}{\sigma}$

$$P(0|0) = \int_{-\infty}^{\frac{-a + \sqrt{E_b}}{\sigma}} \frac{1}{\sqrt{2\pi}} \exp\left[-\frac{z^2}{2}\right] dz \qquad (7.42)$$

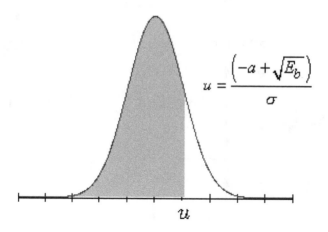

$$u = \frac{\left(-a + \sqrt{E_b}\right)}{\sigma}$$

since $\dfrac{\left(-a + \sqrt{E_b}\right)}{\sigma} > 0$, the shaded area shown is required. The shaded area $= 1 - Q(u)$. Recall that $Q(-u) = 1 - Q(u)$.

Therefore

$$P\left(0|0\right) = 1 - Q\left(\frac{-a + \sqrt{E_b}}{\sigma}\right) \tag{7.43}$$

Similarly

$$P\left(1|0\right) = \int_{-a}^{0} f\left(x_h|0\right) dx_h \tag{7.44}$$

where

$$f\left(x_h|0\right) = \frac{1}{\sigma\sqrt{2\pi}} \exp\left[-\frac{\left(x_h + \sqrt{E_b}\right)^2}{2\sigma^2}\right] \tag{7.45}$$

let $z = \dfrac{\left(x_h + \sqrt{E_b}\right)}{\sigma}$, so that $\dfrac{dz}{dx_h} = \dfrac{1}{\sigma}$

$$P(1|0) = \int_{\frac{-a+\sqrt{E_b}}{\sigma}}^{\frac{\sqrt{E_b}}{\sigma}} \frac{1}{\sqrt{2\pi}} \exp\left[-\frac{z^2}{2}\right] dz \qquad (7.46)$$

Thus

$$P(1|0) = Q\left(\frac{-a+\sqrt{E_b}}{\sigma}\right) - Q\left(\frac{\sqrt{E_b}}{\sigma}\right) \qquad (7.47)$$

(c) The equivalent hard-decision (2-level) DMC is shown below.

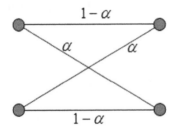

Clearly

$$P(0|0) + P(1|0) = 1 - \alpha \qquad (7.48)$$

because the soft-symbols 0 and 1 are equivalent to a binary digit 0. Thus

$$1 - Q\left(\frac{-a+\sqrt{E_b}}{\sigma}\right) + Q\left(\frac{-a+\sqrt{E_b}}{\sigma}\right) - Q\left(\frac{\sqrt{E_b}}{\sigma}\right) = 1 - \alpha \qquad (7.49)$$

where $\alpha = Q\left(\frac{\sqrt{E_b}}{\sigma}\right)$ and $\sigma = \sqrt{\frac{N_o}{2}}$. Thus, as expected,

$$\alpha = Q\left(\sqrt{\frac{2E_b}{N_o}}\right) \qquad (7.50)$$

12. ▶VIDEO SOLUTION **P12:** $A = \sqrt{\dfrac{2E_s}{T_s}} = 7$ mV, $r_b = 98000$ binary

digits/sec, $M = 8$ $r_s = \dfrac{r_b}{\log_2 M} = \dfrac{r_b}{3} = 32667$ symbols/sec, $T_s = \dfrac{1}{r_s}$ seconds

$E_s = \dfrac{(0.007)^2}{2 * 32667} = 7.5 \times 10^{-10}$ joules. $P_s = 2Q\left(\sqrt{\dfrac{2E_s}{N_o}}\sin\left(\dfrac{\pi}{M}\right)\right) = 5 \times$

10^{-4}. (Using Q-function table). Thus, the number of symbol errors per sec
$= P_s * r_s = 17$ errors/ sec.

- ⭐ SIMULATION **P12:** For further details. Notice how the accuracy
  of the $Q(.)$ function changes the final answer from 17 to 31 symbol errors
  per sec.

13. ▶VIDEO SOLUTION **P13:** (a) For 0 dB and reduction in
throughput, code rate is taken to be 1 and

$$\alpha = Q\left(\sqrt{\dfrac{2E_b}{N_o}}\right) = Q\left(\sqrt{2 * 10^{0/10}}\right) = Q\left(\sqrt{2}\right) = 0.0793 \qquad (7.51)$$

Thus, $P_e = 3\alpha^2(1 - \alpha) + \alpha^3 = 0.018$. (b) For no reduction in throughput

$$\alpha = Q\left(\sqrt{\dfrac{2E_b R_{code}}{N_o}}\right) = Q\left(\sqrt{2 * 10^{0/10} * 1/3}\right) = 0.209 \qquad (7.52)$$

Thus, $P_e = 0.113$.

(c) The repetition code provides a coding gain only if the throughput
is allowed to be reduced.   A reduction in information throughput is not
acceptable in a real communication system. That is, the introduction of error
control coding must reduce the overall P_e without a reduction in throughput.
For example, it is <u>not</u> acceptable to view the 7.00 pm news at 7.05 pm.

14. ▶VIDEO SOLUTION **P14:** (a) Amplitude of signal in general
$A = \sqrt{\dfrac{2E_s}{T_s}}$. For a BPSK signal, $A = \sqrt{\dfrac{2E_s}{T_b}}$. Same amplitude for any

value of M, therefore $\sqrt{\dfrac{2E_s}{T_s}} = \sqrt{\dfrac{2E_b}{T_b}}$, so that $\dfrac{2E_s}{T_s} = \dfrac{2E_b}{T_b}$, and thus $E_s = \dfrac{E_b T_s}{T_b}$.   For no reduction in information throughput, $T_s = (T_b \log_2 M) R_{code}$. Therefore, $E_s = (E_b \log_2 M) R_{code}$.

(b) 1 symbol $= \log_2 M$ binary digits. Therefore, number of binary digits/sec $= 5000 * \log_2 M$. For $M = 2$, 5000 binary digits/sec. For $M = 4$, 10000 binary digits/sec.

(c) Bandwidth required for M-ary PSK
$$B_{MPSK} = \frac{2}{T_s} = \frac{2}{(T_b \log_2 M) R_{code}} = \frac{2r_b}{(\log_2 M) R_{code}}.$$

For $M = 16$, $R_{code} = 1$ and $r_b = 100000$, $B_{MPSK} = \dfrac{200000}{4} = 50000$ Hz. As R_{code} is reduced, B_{MPSK} required is increased.

15. ▶ VIDEO SOLUTION **P15:** Assuming $x_h = \sqrt{E_b}$ if a binary digit is sent under a noiseless channel, the probability receiving a binary digit 0 given that a one is transmitted $P(0|1)$ is the shaded area shown below.

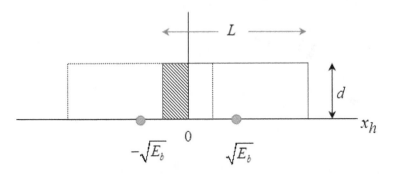

$P(0|1) = $ Shaded area $= \left(\dfrac{L}{2} - \sqrt{E_b}\right) d$

By symmetry, P(1|0) = P(1|0) and the probability of error

$$P_e = P(0|1)P(1) + P(1|0)P(0) = P(0|1)\left[P(1) + P(0)\right] = P(0|1) \qquad (7.53)$$

Thus,

$$P_e = \left(\frac{L}{2} - \sqrt{E_b}\right) d \qquad (7.54)$$

16. ⓥVIDEO SOLUTION **P16:** ⭐SIMULATION **P16:** For the full solution. Experiment with the variables to understand and appreciate their influence on the answer.

17. ⓥVIDEO SOLUTION **P17:** ⭐SIMULATION **P17:** For the full solution. Experiment with the variables to understand and appreciate their influence on the answer.

18. ⓥVIDEO SOLUTION **P18:**

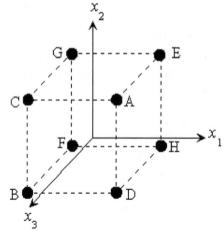

Symbol	Digits
A	111
B	100
C	110
D	101
E	011
F	000
G	010
H	001

For $N = 3$, probability of correctly receiving a symbol $P_c = (1 - \alpha)^3$, where α is the BSC crossover probability. Hence the probability of error in a binary digit, with Gray coding is given by $P_e = \dfrac{1}{3}(1 - P_c) = \dfrac{1}{3}\left(1 - (1 - \alpha)^3\right)$. Thus in general, $P_e = \dfrac{1}{N}\left(1 - (1 - \alpha)^N\right)$, where $\alpha = Q\left(\sqrt{\dfrac{2E_b}{N_o}}\right)$.

19. ⭐ SIMULATION **P19:** For the solution.

20. The distance between neighboring signal points for QAM $= 2A = 2\sqrt{\dfrac{3E_s}{2(M-1)}}$ where $E_s = E_b \log_2 M$. For MPSK, the radius of the circle $r = \sqrt{E_s}$ where the angle $\theta = \dfrac{360}{M}$. Distance between signal points shown in the diagram below is given by the Pythagoras theorem to be $\sqrt{(r - r\cos\theta)^2 + (r\sin\theta)^2}$, where $(r - r\cos\theta)$ is the length of the base and $(r\sin\theta)$ is the height of the triangle drawn by projecting the signal point at angle θ to the horizontal axis.

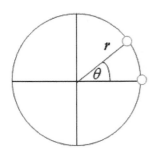

The graph below shows a comparison of the distance between neighboring signal points for $E_b = 1$ J for $M = 4, 16, 32$.

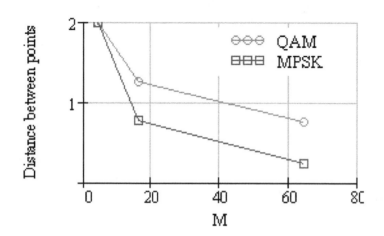

• 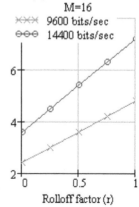 SIMULATION **P20:** For further details.

21. (a) For both MPSK and QAM, the null-to-null bandwidth is given by

$$B = \frac{2}{T_s} = \frac{2}{(T_b \log_2 M) R_{code}} = \frac{2r_b}{(\log_2 M) R_{code}}.$$

With no error-control coding, $R_{code} = 1$ and $B = \dfrac{2r_b}{(\log_2 M)}$.

The second null-to-null bandwidth $= B = 2\left(\dfrac{2}{T_s}\right) = \dfrac{4r_b}{(\log_2 M) R_{code}}$.

(b) The comparison is shown below. (c) See graph below.

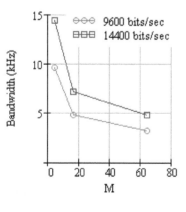

- 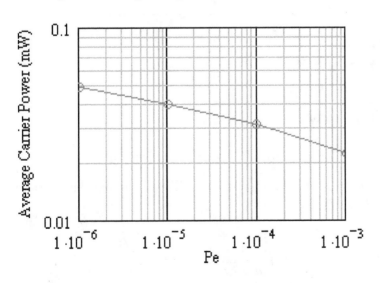SIMULATION **P21:** For further details.

22. For coherent BFSK, $P_e = Q\left(\sqrt{\dfrac{E_b}{N_o}}\right)$.

For

$$P_e = \frac{1}{2\sqrt{\pi}} \exp\left(-\frac{u^2}{2}\right) \tag{7.55}$$

where $u = \sqrt{\dfrac{E_b}{N_o}}$,

$$u^2 = 2\log_e\left(\frac{1}{P_e 2\sqrt{\pi}}\right) = \frac{E_b}{N_o} \tag{7.56}$$

Thus

$$E_b = 2N_o\log_e\left(\frac{1}{P_e 2\sqrt{\pi}}\right) \tag{7.57}$$

and the average carrier power

$$P_{average} = \frac{E_b}{T_b} = E_b r_b = 2N_o r_b \log_e\left(\frac{1}{P_e 2\sqrt{\pi}}\right) \tag{7.58}$$

A graph of $P_{average}$ versus P_e is shown below. As expected, more carrier power is required to improve performance.

- 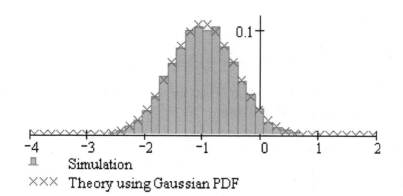 SIMULATION **P22:** For further details.

23. Recall that we require the probability $P(x_h > x_v|0)$ or equivalently, $P(z > 0|0)$ where $z = x_h - x_v$. Transmitting only binary digit zeros, the mean values of the Gaussian random variables x_h and x_v will be 0 and $\sqrt{E_b}$ respectively. Given that the variance of both x_h and x_v is $\dfrac{N_o}{2}$ and that they are statistically independent, the variance of z is equal $= \dfrac{N_o}{2} + \dfrac{N_o}{2} = N_o$ and the mean value of z equal to $(0 - \sqrt{E_b}) = -\sqrt{E_b}$.

- SIMULATION **P23:** Verified by simulation. Refer to the results shown below.

◼   Simulation
×××   Theory using Gaussian PDF

The probability of error

$$P_e = P(z > 0|0) P(0) + P(z < 0|1) P(1) \qquad (7.59)$$

By symmetry, $P(z > 0|0) = P(z < 0|1)$. Thus, the average probability of an error in a binary digit

$$P_e = P(z > 0|0) = \int_0^\infty \frac{1}{\sigma\sqrt{2\pi}} \exp\left[-\frac{(z + \sqrt{E_b})^2}{2\sigma^2}\right] dz \qquad (7.60)$$

where $\sigma = \sqrt{N_o}$. Let $u = \dfrac{z + \sqrt{E_b}}{\sigma}$, then $\dfrac{du}{dz} = \dfrac{1}{\sigma}$. Changing the integration limits, for $z = 0$, $u = \dfrac{\sqrt{E_b}}{\sigma}$ and for $z = \infty$, $u = \infty$. Therefore

$$P_e = \int_{\frac{\sqrt{E_b}}{\sigma}}^{\infty} \frac{1}{\sqrt{2\pi}} \exp\left[-\frac{u^2}{2}\right] du = Q\left(\frac{\sqrt{E_b}}{\sigma}\right) \tag{7.61}$$

$$= Q\left(\sqrt{\frac{E_b}{N_o}}\right)$$

● ⭐ SIMULATION **P23**: For further details.

24. (a) The probability of an error in a binary digit with Gray coding
$$P_e = \frac{M P_s}{2(M-1)} = \frac{64\,(10^{-4})}{2 * 63} = 5.1 \text{ x } 10^{-5}.$$

Note that if it was a MPSK system, then
$$P_e = \frac{P_s}{\log_2 M} = \frac{10^{-4} \log_{10} 2}{\log_{10} 64} = 1.7 \text{ x } 10^{-5}.$$

(b) For coherent orthogonal MFSK,

$$P_s \leq (M-1) Q\left(\sqrt{\frac{E_s}{N_o}}\right) \tag{7.62}$$

For

$$P_s = (M-1) \frac{1}{2\sqrt{\pi}} \exp\left(-\frac{u^2}{2}\right) \tag{7.63}$$

where $u = \sqrt{\dfrac{E_s}{N_o}}$

$$u^2 = 2\log_e\left(\frac{M-1}{P_e 2\sqrt{\pi}}\right) = \frac{E_s}{N_o} \tag{7.64}$$

Thus $E_s = 2N_o \log_e\left(\dfrac{M-1}{P_s 2\sqrt{\pi}}\right) = \dfrac{A^2 T}{2}$. Rearranging this equation,

$$A = \sqrt{\frac{4N_o}{T}\log_e\left(\frac{M-1}{P_s 2\sqrt{\pi}}\right)} = 1.4 \text{ mV} \tag{7.65}$$

☆ **SIMULATION P24:** For the graph shown below. As expected, the amplitude of the carrier has to be increased if T_s is reduced to maintain the energy per symbol and thereby maintain P_s.

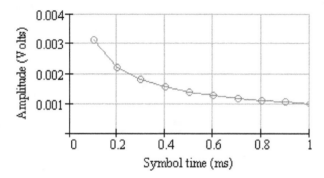

• ☆ **SIMULATION P24:** For further details.

25.

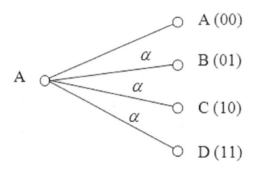

Recall in MFSK, a given signal point can be incorrectly decoded with equal probability as any one of the other $(M-1)$ signal points. Therefore, let α be the forward transition probability $= P(B|A) = P(C|A) = P(D|A)$. Given that

$$P(A|A) + P(B|A) + P(C|A) + P(D|A) = 1 \tag{7.66}$$

we have $P(A|A) = 1 - 3\alpha$. Suppose we transmit N symbols. The average number of binary digits ones received

$$N_{ones} = N_B(1) + N_C(1) + N_D(2) \tag{7.67}$$

where N_B, N_C and N_D are the number of symbols B, C and D received because the symbols B, C, D are replaced by 01, 10 and 11 binary pairs. On average, $\alpha = \dfrac{N_B}{N} = \dfrac{N_C}{N} = \dfrac{N_D}{N}$. Therefore

$$N_{ones} = \alpha N \left(1 + 1 + 2 \right) = 4\alpha N \tag{7.68}$$

The probability of a symbol error

$$
\begin{aligned}
P_s &= \frac{N_B + N_C + N_D}{N} \\
&= \frac{N_B}{N} + \frac{N_C}{N} + \frac{N_D}{N} = \frac{3\alpha N}{N} = 3\alpha
\end{aligned}
\tag{7.69}
$$

Given that the total number of binary digits transmitted is $2N$, the average probability of an error in a binary digit

$$P_e = \frac{N_{ones}}{2N} = \frac{4\alpha N}{2N} = 2\alpha \tag{7.70}$$

Thus

$$P_e = 2\alpha = \frac{2P_s}{3} \tag{7.71}$$

Recall that $N_{ones} = \alpha N \left(1 + 1 + 2 \right) = 4\alpha N$. In general, we would have

$$N_{ones} = \alpha N \left(N_{tup} \right) \tag{7.72}$$

where N_{tup} is the number of binary digit ones within the $n = \log_2 M$ tuple of binary digits. For example, the $n = 2$ tuple is
$\begin{matrix} 0 & 0 \\ 0 & 1 \\ 1 & 0 \\ 1 & 1 \end{matrix}$
in which $N_{tup} = 4$

and the $n = 3$ tuple is
$\begin{matrix} 0 & 0 & 0 \\ 0 & 0 & 1 \\ 0 & 1 & 0 \\ 0 & 1 & 1 \\ 1 & 0 & 0 \\ 1 & 0 & 1 \\ 1 & 1 & 0 \\ 1 & 1 & 1 \end{matrix}$
in which $N_{tup} = 12$. In general, we have

$N_{tup} = \dfrac{nM}{2}$ where $M = 2^n$.

Given that n binary digits correspond to N symbols, we have

$$P_e = \frac{N_{ones}}{nN} = \frac{\alpha N N_{tup}}{nN} = \frac{\alpha N_{tup}}{n} = \frac{\alpha n M}{n2} = \frac{\alpha M}{2} \tag{7.73}$$

Recall that for $n = 2$, we had $P_s = \frac{N_B}{N} + \frac{N_C}{N} + \frac{N_D}{N} = \frac{3\alpha N}{N}$. Thus in general

$$P_s = \frac{\left(2^n - 1 \right) \alpha N}{N} = \alpha \left(2^n - 1 \right) \tag{7.74}$$

But $M = 2^n$ so that we may write $P_s = \alpha \left(M - 1 \right)$ or $\alpha = \frac{P_s}{M-1}$ so that finally, we have

$$P_e = \frac{M\alpha}{2} = \frac{M P_s}{2 \left(M - 1 \right)} \tag{7.75}$$

26. Consider the noise component output n_1 which is given by

$$n_1 = \int_0^{T_b} n(t)\phi(t)dt \tag{7.76}$$

$$E\left[n_1\right] = E\left[\int_0^{T_b} n(t)\phi(t)dt\right] = \int_0^{T_b} E\left[n(t)\right]\phi(t)dt \qquad (7.77)$$

but $E\left[n(t)\right] = 0$ (zero mean) and thus $E\left[n_1\right] = 0$. The variance

$$\sigma^2 = E\left[n_1^2\right] = E\left[\left[\int_0^{T_b} n(t)\phi(t)dt\right]^2\right] \qquad (7.78)$$

$$= \int_0^{T_b}\int_0^{T_b} E\left[n(t)n(\tau)\right]\phi(t)\phi(\tau)dtd\tau$$

But

$$E\left[n(t)n(\tau)\right] = R_X(t-\tau) = \frac{N_o}{2}\delta(t-\tau) \qquad (7.79)$$

Therefore

$$\sigma^2 = \int_0^{T_b}\int_0^{T_b} \frac{N_o}{2}\delta(t-\tau)\phi(t)\phi(\tau)dtd\tau \qquad (7.80)$$

But

$$\int_0^{T_b} \frac{N_o}{2}\delta(t-\tau)dt = \frac{N_o}{2} \qquad (7.81)$$

Hence

$$\sigma^2 = \frac{N_o}{2}\int_0^{T_b} \phi^2(\tau)d\tau = \frac{N_o}{2} \qquad (7.82)$$

since the energy of the basis function $\phi(t)$ is 1 joule. By symmetry, $E\left[n_2\right] = 0$ and $\sigma^2 = E\left[n_2^2\right] = \frac{N_o}{2}$.

27. (a) The QPSK signal constellation diagram is shown below where $0(-1,-i)$, $1(1,-i)$, $2(-1,+i)$, $3(1,+i)$.

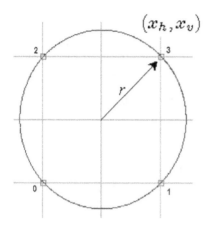

The radius $r = \sqrt{E_s}$ where E_s is energy per QPSK signal.   Now $r = \sqrt{1^2 + 1^2} = \sqrt{2}$, so that $E_s = 2$ Joules and for uncoded QPSK, $E_s = 2E_b$, so that $E_b = 1$ Joule.

(b) For $X_k = \begin{bmatrix} 1 - i \\ -1 + i \\ 1 - i \\ 1 + i \end{bmatrix}$, the signal points transmitted are 1, 2, 1, 3.

Now $Euler(n, k) = e^{j\frac{\pi n k}{2}}$ for which

$$Euler\,(2, k) = \begin{bmatrix} 1 \\ -1 \\ 1 \\ -1 \end{bmatrix} \tag{7.83}$$

for $k = 0 \cdots 3$. We know that

$$x_n = \frac{1}{\sqrt{N}} \sum_{k=0}^{N-1} X_k e^{j\frac{(2\pi n k)}{N}} \tag{7.84}$$

for $n = 0, 1, \cdots N - 1$ which for $n = 0$ and $N = 4$ is
x

$$0 = \frac{1}{\sqrt{4}} \sum_{k=0}^{3} X_k = \frac{1}{2}(1 - i - 1 + i + 1 - i + 1 + i) = \frac{1}{2}(2) = 1 \tag{7.85}$$

Therefore the first sample point is $x_0\sqrt{N} = 1\sqrt{4} = 2$ volts as expected. Now for $n = 2$,

$$x_2 = \frac{1}{\sqrt{N}}\sum_{k=0}^{3}X_k e^{j\pi k} \tag{7.86}$$

$$= \frac{1}{2}(X_0(1) + X_1(-1) + X_2(1) + X_3(-1))$$

$$= \frac{1}{2}(X_0 - X_1 + X_2 - X_3)$$

$$= \frac{1}{2}(1 - i + 1 - i + 1 - i - 1 - i) = \frac{1}{2}(2 - 4i) = 1 - 2i$$

Therefore the second sample point is $x_2\sqrt{4} = 2 - 4i$ and the real part is 2 volts as expected.

28. (a) The signal points $p_0 = e^{j\pi} = -1$ and $p_1 = e^{j0} = 1$ so that

$$X_k = \begin{bmatrix} -1 \\ 1 \end{bmatrix}. \text{ Thus } X'_k = \begin{bmatrix} 0 \\ p_0 \\ p_1 \\ 0 \\ p_1^* \\ p_0^* \end{bmatrix} = \begin{bmatrix} 0 \\ -1 \\ 1 \\ 0 \\ 1 \\ -1 \end{bmatrix}.$$

Given that $K = 2$, we have

$$g'(t) = \sum_{k=0}^{K-1}X_k e^{j\frac{2\pi(k+1)t}{T}} = \sum_{k=0}^{1}X_k e^{j\frac{2\pi(k+1)t}{T}} \tag{7.87}$$

$$= X_0 e^{j\frac{2\pi t}{T}} + X_1 e^{j\frac{4\pi t}{T}}$$

and

$$v_{bandband}(t) = \text{Re}\left(g'(t)\right)$$

$$= \text{Re}\left(-e^{j\frac{2\pi t}{T}}\right) + \text{Re}\left(e^{j\frac{4\pi t}{T}}\right) \tag{7.88}$$

$$= -\cos\left(\frac{2\pi t}{T}\right) + \cos\left(\frac{4\pi t}{T}\right)$$

where the spectral component $-\cos\left(\frac{2\pi t}{T}\right)$ represents the binary digit '0' with frequency $\frac{1}{T}$ Hz and the spectral component $\cos\left(\frac{4\pi t}{T}\right)$ represents the binary digit '1' with frequency $\frac{2}{T}$ Hz. The separation is $\frac{2}{T} - \frac{1}{T} = 1/T$ Hz as expected.

(b) For $n = 1$ and $N = 6$

$$x_1 = \frac{1}{\sqrt{6}}\sum_{k=0}^{5} X_k' e^{j\frac{2\pi k}{6}} \tag{7.89}$$

$$= \frac{1}{\sqrt{6}}\left(X_0' + X_1' e^{j\frac{2\pi}{6}} + X_2' e^{j\frac{4\pi}{6}} + X_3' e^{j\frac{6\pi}{6}} + X_4' e^{j\frac{8\pi}{6}} + X_5' e^{j\frac{10\pi}{6}}\right)$$

$$= \frac{1}{\sqrt{6}}\left(-e^{j\frac{2\pi}{6}} + e^{j\frac{4\pi}{6}} + e^{j\frac{8\pi}{6}} - e^{j\frac{10\pi}{6}}\right) = \frac{-2}{\sqrt{6}}$$

so that the second sample point $x_1 \frac{\sqrt{N}}{2} = \frac{-2}{\sqrt{6}}\frac{\sqrt{6}}{2} = -1$ as expected.

(c) Recover via the DFT of y_n i.e.

$$Y_k = \frac{1}{\sqrt{6}}\sum_{k=0}^{5} y_n e^{-j\frac{(2\pi n k)}{6}} \tag{7.90}$$

where $\{y_n\}$ are the received samples of the transmitted $\{x_n\}$ samples and $k = 0, 1, \cdots, K$ with the demodulated complex points within $Y_1, Y_2, \cdots, Y_K$ and $Y_0 = 0$.

(d) As K is increased, the frequency of the subcarriers ranges from $\frac{1}{T}$, $\frac{2}{T}, ..., \frac{K}{T}$, with the highest spectral component at $\frac{K}{T}$ Hz. The null bandwidth can be taken to be $B_{null} = \frac{K+1}{T} = \frac{K+1}{KT_s}$ and for $K \gg 1$, $B_{null} \approx \frac{K}{KT_s} = \frac{1}{T_s}$ where $T_s = T_b \log_2 M$. For BPSK, $M = 2$ and thus $T_s = T_b \log_2 2 = T_b$. Thus $B_{null} = \frac{1}{T_b}$. The implication is that the bandwidth of the OFDM signal is essentially as the same as the bandwidth of the BPSK signal and essentially independent of K! The penalty incurred is that the subcarriers are all closer spaced together (separation of $\frac{1}{T} = \frac{1}{KT_s}$) as K is increased which also increases the complexity of the communication system.

29. (a) Recall that since 4-QAM is equivalent to QPSK, the

$$v_{baseband}(t) = \mathrm{Re}\,(g(t)) = \sum_{k=0}^{K-1} r \cos\left(\frac{2\pi kt}{T} + \theta_k\right) \qquad (7.91)$$

from which the amplitude of a spectral component is $r = \sqrt{E_s} = \sqrt{A^2\,(a_i^2 + b_i^2)} = \sqrt{2}$ for $A = 1$ with a_i and b_i from the set $\in \{\pm 1\}$. A sketch of the single-sided spectrum is shown below.

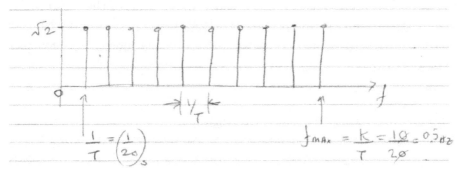

(b) The highest frequency is $\frac{K}{T} = \frac{K}{KT_s} = \frac{1}{T_s} = 0.5$ Hz. The bandwidth of the OFDM signal can be taken to be the null bandwidth $B_{null} = \frac{K+1}{T}$. For large K, we can approximate $B_{null} \approx \frac{K}{KT_s} = \frac{1}{T_s} = 0.5$ Hz.

(c) The bandwidth of the OFDM signal be reduced by increasing the time duration T_s.

(d) As K is increased, the bandwidth remains unaffected (assuming T_s does not change) as the spectral components come closer together because the separation between the subcarriers is $\frac{1}{KT_s}$ Hz.

30. (a) Given $K = 4$ and the symbols sent are 0, 7, 7, 10. Thus $N = 2\,(K+1) = 2\,(4+1) = 10$.

$$X_k = \begin{bmatrix} -3+3i \\ -1-3i \\ -1-3i \\ 1-i \end{bmatrix} \quad \text{and} \quad X'_k = \begin{bmatrix} 0 \\ -3+3i \\ -1-3i \\ -1-3i \\ 1-i \\ 0 \\ 1+i \\ -1+3i \\ -1+3i \\ -3-3i \end{bmatrix} \qquad (7.92)$$

(b) The $x_n \frac{\sqrt{N}}{2}$ values correspond to the real samples of the OFDM signal, where x_n is determined from the IDFT of $\{X'_k\}$.

(c) (i) The frequency resolution of the FFT is the separation of the spectral lines shown below, given by $\frac{0.2}{4} = \frac{2}{40} = \frac{1}{20}$ Hz.

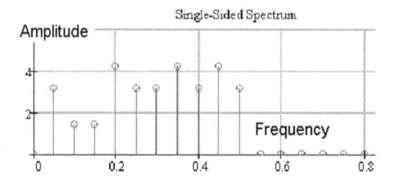

Single-Sided Spectrum

(ii) In OFDM, the subcarriers are separated in frequency by $\frac{1}{T}$ Hz, where $T = KT_s$ is the OFDM signal duration.   Given that there is no smearing effect in the FFT spectrum, the frequency resolution of the FFT $\frac{1}{20} = \frac{1}{T}$, so that $T = 20$ seconds.   (iii) The number of real samples $N = 2(K+1) = 2(10+1) = 22$. (iv) The time duration of a QAM signal $T_s = \frac{T}{K} = 2$ seconds. Its interesting to note that T_s is the time duration of the OFDM signal if only a single symbol is transmitted. (v) The highest frequency within this OFDM signal is $\frac{K}{T} = \frac{10}{20} = 0.5$ Hz which is also evident from the single-sided spectrum. (vi) The bandwidth B_{null} of this OFDM signal as K is increased is approximately given by $\frac{1}{T_s} = \frac{1}{2} = 0.5$ Hz.

31. (a) For MPSK, as M is increased, the bandwidth efficiency improved. The penalty incurred is the SNR has to be icreased to maintain P_e. This is because the signal points, which lie on a circle of radius $\sqrt{E_s}$, come closer together as M is increased, which in turn would increase P_s if SNR was not increased to compensate. MPSK is suitable for band-limited channels where we desire $\frac{r_b}{B} > 1$.

(b) QAM is overall better than MPSK because for a given bandwidth efficiency, the required increase in SNR to maintain P_e is less. This is because the QAM signal space is rectangular and for a given signal power, the QAM signal points are further apart than the corresponding MPSK signal points. It should be noted however that the unlike MPSK, the amplitude of the carrier in QAM does not remain constant, which for certain types of noise can make it unsuitable!

(c) For MFSK, as M is increased, the bandwidth efficiency decreases. Advantage is a reduction in SNR for a given P_e because of the M-dimensional signal space. MFSK is suitable for power-limited channels.

(d) (i) Using gray coding, $P_e = \frac{MP_s}{2(M-1)} = \frac{64(10^{-4})}{2(63)} = 5.1 \times 10^{-6}$. (ii) Consider the limit where the probability of a symbol error

$$P_s = (M-1)Q\left(\sqrt{\frac{E_s}{N_o}}\right) = (M-1)\frac{1}{2\sqrt{\pi}}\exp\left(-\frac{u^2}{2}\right) \qquad (7.93)$$

where $u = \sqrt{\frac{E_s}{N_o}}$

$$u^2 = 2\log_e\left(\frac{M-1}{P_s 2\sqrt{\pi}}\right) = \frac{E_s}{N_o} \qquad (7.94)$$

Therefore

$$E_s = 2N_o\log_e\left(\frac{M-1}{P_s 2\sqrt{\pi}}\right) = \frac{A^2 T_s}{2} \qquad (7.95)$$

so that

$$A = \sqrt{\frac{4N_o}{T_s}\log_e\left(\frac{M-1}{P_s 2\sqrt{\pi}}\right)} = 2.4 \text{ mV} \qquad (7.96)$$

32. For uncoded QPSK, the approximate null bandwidth $B = \frac{2}{T_s} = \frac{2}{T_b \log_2 4} = \frac{1}{T_b}$ allowing a data rate $r_b = \frac{1}{T_b} = B$. However, using OFDM with QPSK subcarrriers, $B = \frac{1}{T_s} = \frac{1}{T_b \log_2 4}$ which now allows a date rate of $r_b = \frac{1}{T_b} = 2B$, which is better by a factor of two.

Chapter 8

Block and Convolutional Codes

8.1 Solutions

1. ▶VIDEO SOLUTION **P1:** ★SIMULATION **P1:** For the full solution.

2. ★SIMULATION **P2:** For the full solution.

3. ★SIMULATION **P3:** For the full solution.

4. ★SIMULATION **P4:** For the full solution.

5. ★SIMULATION **P5:** For the full solution.

6. (a) ★SIMULATION **P6:** For the full solution.

(b) (i) From the generator matrix

$$\mathbf{G} = \begin{bmatrix} 1 & 1 & 0 & 1 & 0 \\ 1 & 0 & 1 & 0 & 1 \end{bmatrix} \tag{8.1}$$

the number of rows is $k = 2$ and the number of columns is $n = 5$. Thus,

for all possible inputs $\left\{ \begin{array}{cc} 0 & 0 \\ 0 & 1 \\ 1 & 0 \\ 1 & 1 \end{array} \right\}$, the code words are

$$\mathbf{c_0} = \mathbf{uG} = \begin{bmatrix} 0 & 0 \end{bmatrix} \begin{bmatrix} 1 & 1 & 0 & 1 & 0 \\ 1 & 0 & 1 & 0 & 1 \end{bmatrix} = \begin{bmatrix} 0 & 0 & 0 & 0 & 0 \end{bmatrix} \text{ weight } \mathbf{w_0} = 0$$

$$\mathbf{c_1} = \mathbf{uG} = \begin{bmatrix} 0 & 1 \end{bmatrix} \begin{bmatrix} 1 & 1 & 0 & 1 & 0 \\ 1 & 0 & 1 & 0 & 1 \end{bmatrix} = \begin{bmatrix} 1 & 0 & 1 & 0 & 1 \end{bmatrix} \text{ weight } \mathbf{w_1} = 3$$

$$\mathbf{c_2} = \mathbf{uG} = \begin{bmatrix} 1 & 0 \end{bmatrix} \begin{bmatrix} 1 & 1 & 0 & 1 & 0 \\ 1 & 0 & 1 & 0 & 1 \end{bmatrix} = \begin{bmatrix} 1 & 1 & 0 & 1 & 0 \end{bmatrix} \text{ weight } \mathbf{w_2} = 3$$

$$\mathbf{c_3} = \mathbf{uG} = \begin{bmatrix} 1 & 1 \end{bmatrix} \begin{bmatrix} 1 & 1 & 0 & 1 & 0 \\ 1 & 0 & 1 & 0 & 1 \end{bmatrix} = \begin{bmatrix} 0 & 1 & 1 & 1 & 1 \end{bmatrix} \text{ weight } \mathbf{w_3} = 4$$

From which the minimum weight is 3 and hence the error-control capability is $\left\lfloor \frac{3-1}{2} \right\rfloor = 1$.

(ii) Selecting say $\mathbf{c_1}$ and $\mathbf{c_2}$, we get

$$\begin{array}{ccccccc} & 1 & 0 & 1 & 0 & 1 & \quad \mathbf{c_1} \\ + & 1 & 1 & 0 & 1 & 0 & \quad \mathbf{c_2} \\ \hline & 0 & 1 & 1 & 1 & 1 & \quad \mathbf{c_1} + \mathbf{c_2} = \mathbf{c_3} \end{array}$$

(iii) The generator matrix

$$\mathbf{G} = \begin{bmatrix} 1 & 1 & 0 & 1 & 0 \\ 1 & 0 & 1 & 0 & 1 \end{bmatrix} = \begin{bmatrix} \mathbf{P} & \mathbf{I_2} \end{bmatrix} \tag{8.2}$$

so that

$$\mathbf{P} = \begin{bmatrix} 1 & 1 & 0 \\ 1 & 0 & 1 \end{bmatrix} \tag{8.3}$$

$$\mathbf{P}^T = \begin{bmatrix} 1 & 1 \\ 1 & 0 \\ 0 & 1 \end{bmatrix}$$

hence

$$\mathbf{H} = \begin{bmatrix} \mathbf{I}_{n-k} & \mathbf{P}^T \end{bmatrix} = \begin{bmatrix} 1 & 0 & 0 & 1 & 1 \\ 0 & 1 & 0 & 1 & 0 \\ 0 & 0 & 1 & 0 & 1 \end{bmatrix} \tag{8.4}$$

Finally to check, we find that

$$\mathbf{GH}^T = \begin{bmatrix} 1 & 1 & 0 & 1 & 0 \\ 1 & 0 & 1 & 0 & 1 \end{bmatrix} \begin{bmatrix} 1 & 0 & 0 \\ 0 & 1 & 0 \\ 0 & 0 & 1 \\ 1 & 1 & 0 \\ 1 & 0 & 1 \end{bmatrix} \tag{8.5}$$

$$= \begin{bmatrix} 0 & 0 & 0 \\ 0 & 0 & 0 \end{bmatrix}$$

as expected.

7. ▶VIDEO SOLUTION **P7**: ★SIMULATION **P7**: For the full solution.

8. (a) For $m = 4$, $n = 16 - 1 = 15$ and $k = 16 - 1 - 4 = 11$ and the 4-bit tuple is

```
0 0 0 0
0 0 0 1
0 0 1 0
0 0 1 1
0 1 0 0
0 1 0 1
0 1 1 0
0 1 1 1
1 0 0 0
1 0 0 1
1 0 1 0
1 0 1 1
1 1 0 0
1 1 0 1
1 1 1 0
1 1 1 1
```

Now we require

$$\mathbf{H} = \begin{bmatrix} \mathbf{P}^T & \mathbf{I}_{n-k} \end{bmatrix} \tag{8.6}$$

where

$$n - k = 15 - 11 = 4 \tag{8.7}$$

Therefore, the generator matrix is of the form $\mathbf{G} = \begin{bmatrix} \mathbf{I}_k & \mathbf{P} \end{bmatrix}$ to ensure that $\mathbf{GH}^T = \mathbf{0}$. Using the 4-bit tuple table and **not** making use of 0000 while ensuring the last 4 columns correspond to the identity matrix $\mathbf{I}_4$:

$$\mathbf{H} = \begin{bmatrix} 0 & 0 & 0 & 0 & 1 & 1 & 1 & 1 & 1 & 1 & 1 & 1 & 0 & 0 & 0 \\ 0 & 1 & 1 & 1 & 0 & 0 & 0 & 1 & 1 & 1 & 1 & 0 & 1 & 0 & 0 \\ 1 & 0 & 1 & 1 & 0 & 1 & 1 & 0 & 0 & 1 & 1 & 0 & 0 & 1 & 0 \\ 1 & 1 & 0 & 1 & 1 & 0 & 1 & 0 & 1 & 0 & 1 & 0 & 0 & 0 & 1 \end{bmatrix} \tag{8.8}$$

(b) ⭐ SIMULATION **P8:** For the full solution.

9. ▶ VIDEO SOLUTION **P9:** ⭐ SIMULATION **P9:** For the full solution.

10. ⭐ SIMULATION **P10:** For the full solution.

11. ⭐ SIMULATION **P11:** For the full solution.

12. ▶ VIDEO SOLUTION **P12:** (a) $(X^7 + 1)$ divided by $g(X) = X^4 + X^2 + X + 1$ should leave a remainder of zero

```
                              1   0   1   1
    1  0  1  1  1 | 1   0   0   0   0   0   0   1
                    1   0   1   1   1
                    ─────────────────────
                        1   1   1   0   0   1
                        1   0   1   1   1
                        ─────────────────────
                            1   0   1   1   1
                            1   0   1   1   1
                            ─────────────────────
                            0   0   0   0   0
```

(b)

$$m(X) = X^2 \equiv 100 \tag{8.9}$$

$$X^{n-k}m(X) = X^4 m(X) = X^6 \equiv 1000000 \tag{8.10}$$

Now we divide $X^{n-k}m(X) \equiv 1000000$ by $g(X) \equiv 10111$ as follows.

```
                    1  0  1   a(X)
1 0 1 1 1 | 1  0  0  0  0  0  0
            1  0  1  1  1
            ───────────────
               1  1  1  0  0
               1  0  1  1  1
               ───────────────
                  1  0  1  1   p(X)
```

Therefore the code word is given by

```
      1  0  0  0  0  0  0   X^{n-k}m(X)
+           1  0  1  1      p(X)
      ─────────────────────
      1  0  0  1  0  1  1   c(X)
```

(c)

$$\begin{aligned}
c(X) &= a(X)g(X) = \left(X^2 + 1\right)\left(X^4 + X^2 + X + 1\right) \tag{8.11}\\
&= X^6 + X^4 + X^3 + X^2 + X^4 + X^2 + X + 1\\
&= X^6 + X^3 + X + 1\\
&\equiv 1001011
\end{aligned}$$

13. (a)

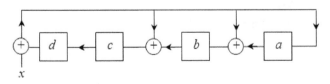

- ⭐ SIMULATION **P13:** For the rest of the solution.

14. ⭐ SIMULATION **P14:** For the full solution.

15. ⭐ SIMULATION **P15:** For the full solution.

16. ▶ VIDEO SOLUTION **P16:** (a) $G_1^{(1)}(D) = 1+D$ and $G_1^{(2)}(D) = 1$
Thus

$$\begin{aligned} G(D) &= \left[\begin{array}{cc} G_1^{(1)}(D) & G_1^{(2)}(D) \end{array} \right] \\ &= \left[\begin{array}{cc} 1+D & 1 \end{array} \right] \end{aligned} \tag{8.12}$$

(b) The message sequence $m(D) = 1 + D^2 + D^3 + D^4 + \ldots$

$$\begin{aligned} x^{(1)}(D) &= m(D)G_1^{(1)}(D) \\ &= \left(1 + D^2 + D^3 + D^4 + \ldots\right)(1+D) \\ &= 1 + D^2 + D^3 + D^4 + \ldots + D + D^3 + D^4 + \ldots \\ &= 1 + D + D^2 \end{aligned} \tag{8.13}$$

or equivalently 11100000000000...

$$\begin{aligned} x^{(2)}(D) &= m(D)G_1^{(2)}(D) = \left(1 + D^2 + D^3 + D^4 + \ldots\right)(1) \\ &= 1 + D^2 + D^3 + D^4 + \ldots \end{aligned} \tag{8.14}$$

or equivalently 1011111111....
Multiplexing the two sequences shown in the below

11100000000000...
10111111111111....

The output of the encoder is 11, 10, 11, 01, 01, 01, 01, ...

(c)

Input	Initial State	Final State	Code Word
0	0	0	00
1	0	1	11
0	1	0	10
1	1	1	01

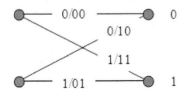

(d) If the initial state cannot be assumed to be the all-zero state, the metric of each state is set to be zero. Code word sequence received is 11, 10, 11, 01, 01, 01, 01, ...

(Trellis split into two to show within a page width)

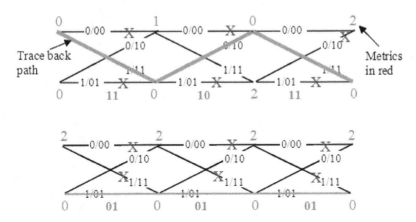

Decoded bits from trace back path = 101111

(e)

Encoder Output	11, 10, 11, 01, 01, 01, 01, 01, 01, ...
Channel Output	01, 10, 01, 01, 11, 01, 11, 01, 11, ...

(Trellis split into two to show within a page width)

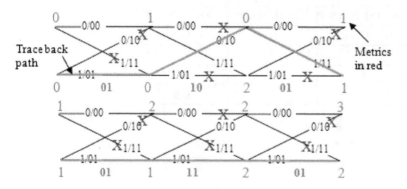

Decoded bits from trace back path = 101111

(f) Draw a trellis diagram of the code. Starting in the all-zero state, follow the branch which deviates from this all-zero state. Make a note of this new state. Now find the minimum weight path back to the all-zero state. Free distance = minimum weight path which deviated from the all-zero state and returned to the all-zero state. This is shown in blue in the diagram below. Hence free distance of this code = 3.

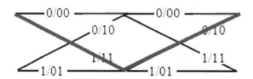

17. (a)

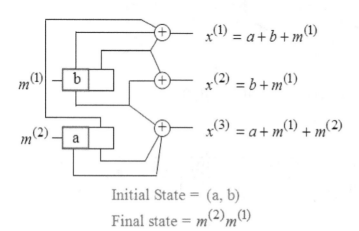

$$x^{(1)} = a + b + m^{(1)}$$

$$x^{(2)} = b + m^{(1)}$$

$$x^{(3)} = a + m^{(1)} + m^{(2)}$$

Initial State = (a, b)

Final state = $m^{(2)} m^{(1)}$

Input $m^{(1)}m^{(2)}$	Initial State (a, b)	Final State $m^{(2)}m^{(1)}$	Code word $x^{(1)}x^{(2)}x^{(3)}$
00	00	00	000
01	00	10	001
10	00	01	111
11	00	11	110
00	01	00	110
01	01	10	111
10	01	01	001
11	01	11	000

Input $m^{(1)}m^{(2)}$	Initial State (a, b)	Final State	Code word $x^{(1)}x^{(2)}x^{(3)}$
00	10	00	101
01	10	10	100
10	10	01	010
11	10	11	011
00	11	00	011
01	11	10	010
10	11	01	100
11	11	11	101

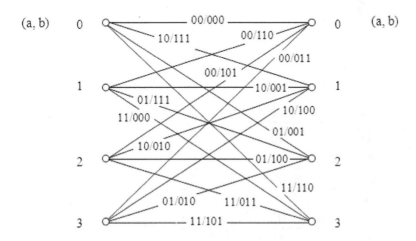

(b)

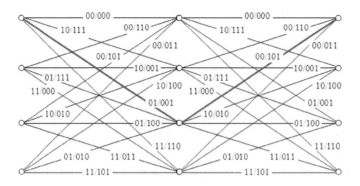

Hence the free distance is the weight of the code word path shown high-lighted, which is equal to 3.

18. (a)

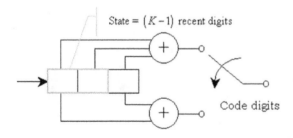

State $= (K-1)$ recent digits

Code digits

(b)

Input	Initial State	Final State	Code word
0	00	00	00
1	00	10	11
0	01	00	11
1	01	10	00
0	10	01	10
1	10	11	01
0	11	01	01
1	11	11	10

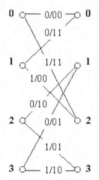

(c)

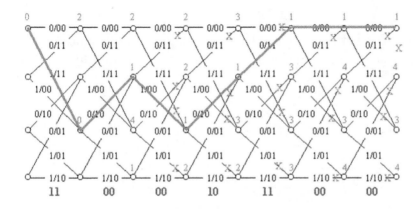

Decoded information sequence	1010000
Received from channel	11 00 00 10 11 00 00
Code words on trace back path	11 10 00 10 11 00 00

Channel error

19. (a)

Symbol	Code word	Hamming Distance from code word 111010
A	000000	4
B	101010	1
C	010101	5
D	111111	2

Since a maximum likelihood decoder is used, the symbol with the minimum hamming distance will be output by the decoder = symbol B.

Transmitted	101010
Received	111010

Probability $P(111010 \mid 10101) = (1 - \alpha)^5 \alpha = 0.06$.

20. (a) The Information throughput is reduced by a factor of R_{code}. This is not acceptable.

(b) The solution is to reduce the time duration per code symbol such that the code digits occupy the same time space as the corresponding information digit(s).

21. (a) Using the state diagram and starting the all-zero state, we find

Input	1, 0, 1, 0, 0	Two zeros added to flush the encoder
Output	11, 10, 00, 10, 11	

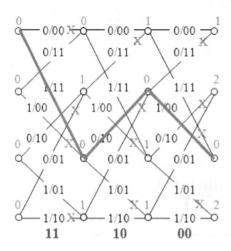

Assumed the encoder starts in the all-zero state.

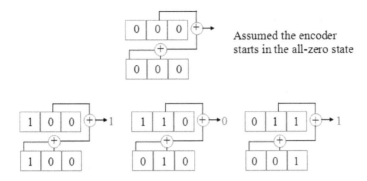

The configuration is the inverse operation of the encoding process.

22. (a) Shannon's coding theorem $r_b\Omega(p) \le r_cC$, where $\Omega(p)$ is the source entropy of the binary source (bits/symbol), $r_b\Omega(p)$ is the information rate from the binary source (bits/sec), C is the maximum average mutual information (bits/symbol). Since $R_{code} = \dfrac{r_b}{r_c}$, $R_{code}\Omega(p) \le C$. We can control p to make $\Omega(p) \approx 1$ using an appropriate source encoder, thus $R_{code} \le C$. The implication is that there is an error-control code with a code rate $R_{code} \le C$, which can provide an arbitrarily low probability of error P_e. The surprising result is that we do **not** need $R_{code} \ll C$ to achieve a low P_e.

(b)

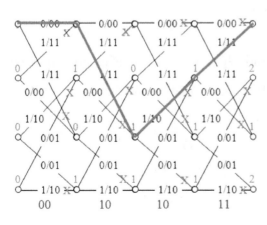

Code digit corrupted

00	1̶0̶	10	11	Received sequence
00	11	10	11	Decoded codeword sequence
0	1	1	1	Information sequence

23.

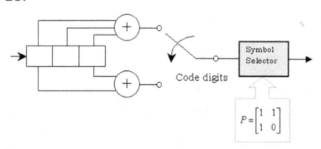

To determine the generator transfer function matrix of the punctured convolutional code, first we extract impulse response by inserting a binary 1 followed by an all-zero sequence. From the puncturing matrix, we know that the code rate is 2/3. Hence input the sequence 10000... and 01000... and segment the output in blocks of three.

Input	Output
10000...	111, 110, 000, ...
01000...	001, 101, 000, ...

Therefore $G_0 = \begin{bmatrix} 1 & 1 & 1 \\ 0 & 0 & 1 \end{bmatrix}$ and $G_1 = \begin{bmatrix} 1 & 1 & 0 \\ 1 & 0 & 1 \end{bmatrix}$ Multiply the elements of G_1 by D and add to the elements of G_0 to write down the generator transfer function matrix as follows:

$$G(D) = \begin{bmatrix} 1+D & 1+D & 1 \\ D & 0 & 1+D \end{bmatrix} \qquad (8.15)$$

24. (a)

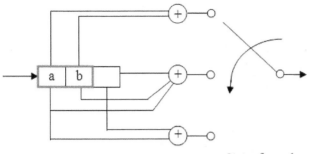

State of encoder = (a, b)

(b)

Input	Initial State	Final State	Output
0	00	00	000
1	00	10	111
0	01	00	011
1	01	10	100
0	10	01	110
1	10	11	001
0	11	01	101
1	11	11	010

(c) Free distance is equal to the weight on the path highlighted, which is 7.

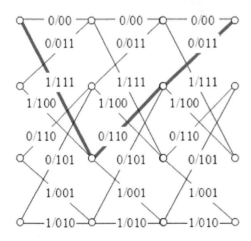

25. Signal energy

$$\begin{aligned} E_s(TCM) &= R_{code}\, E_b \log_2 M \qquad\qquad (8.16)\\ &= \frac{2}{3} E_b \log_2 8 = 2E_b \end{aligned}$$

Thus radius of 8-PSK signal constellation is

$$\sqrt{E_s} = \sqrt{2E_b} \qquad\qquad (8.17)$$

$$E_s(QPSK) = R_{code}\, E_b \log_2 M = 1 * E_b \log_2 4 = 2E_b \qquad (8.18)$$

Thus radius of QPSK signal constellation is also $\sqrt{2E_b}$. Recall d_{free} is the Euclidian distance between signal point 4 and 0, which is the diameter of the circle. Thus, $d_{free} = 2\sqrt{2E_b}$. Similarly, d_{ref} in this case is simply the distance between adjacent signal points on a QPSK signal constellation in which the radius of the circle is $\sqrt{2E_b}$.

i.e.

$$\begin{aligned} x^2 &= r^2 + r^2 = 2r^2 \qquad\qquad (8.19)\\ &= 2r^2 = 2\,(2E_b) = 4E_b\\ x &= \sqrt{4E_b} = d_{ref} \end{aligned}$$

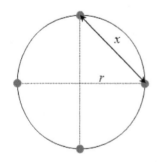

The asymptotic coding gain

$$G(dB) = 10 \log_{10} \left(\frac{d_{free}^2}{d_{ref}^2} \right) \tag{8.20}$$

$$= 10 \log_{10} \left(\frac{4 (2E_b)}{4E_b} \right) = 10 \log_{10}(2) = 3 \text{ dB}$$

as expected.

26. (a) From the table we note that for an adjacent symbol error e.g 4 (transmitted) and received as either symbol 3 or 5, the number of binary digit errors = 1.

Symbol	Mapping
4	110
3	010
5	111

(b) Without loss of generality, we can assume the radius of the circle = $r = \sqrt{E_s} = 1$.

$$x = r \sin \theta \tag{8.21}$$
$$d_1 = 2x = 2r \sin \theta = 2r \sin (22.5) = 0.77$$
$$d_2 = 2r \sin 45 = \sqrt{2} = 1.41$$
$$d_3 = 2r \sin (67.5) = 1.85$$
$$d_4 = 2$$

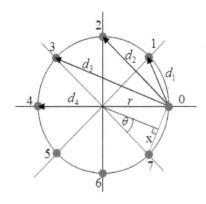

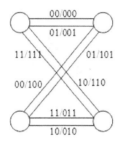

(c)

Input (m_1)	Initial State	Final State	Output code word ($v_1 v_2$)
0	0	0	00
1	0	1	11
0	1	0	10
1	1	1	01

Since $v_3 = m_2$, a branch code word $m_1 m_2 / (v_1 v_2 v_3)$

(d)

Input	10, 00, 00
Code word	110, 100, 000
Signal point	4, 7, 0

Codeword	Signal Point
000	0
001	1
011	2
010	3
110	4
111	5
101	6
100	7

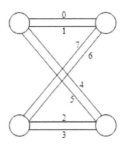

(e) Received signal points (2, 7, 0)

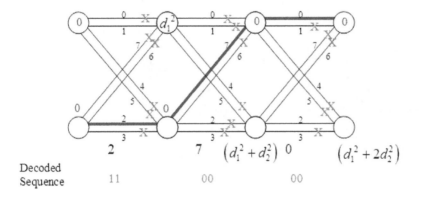

Decoded
Sequence   11   00   00

27.

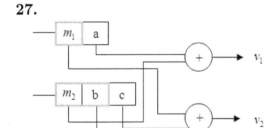

The diagram shows the contents of the shift registers after the input of $m_1 m_2$, assuming the initial state of the encoder is (abc). Thus

$$v_1 = a + m_2 \qquad (8.22)$$
$$v_2 = m_1 + c$$
$$v_3 = b$$

Input $(m_1 m_2)$	Initial State (abc)	Final State	Output code word $(v_1 v_2 v_3)$
00	000	000	000
01	000	010	100
10	000	100	010
11	000	110	110
00	001	000	010
01	001	010	110
10	001	100	000
11	001	110	100
00	010	001	001
01	010	011	101
10	010	101	011
11	010	111	111
00	011	001	011
01	011	011	111
10	011	101	001
11	011	111	101

$$\begin{aligned} v_1 &= a + m_2 \\ v_2 &= m_1 + c \\ v_3 &= b \end{aligned} \tag{8.23}$$

Input (m_1m_2)	Initial State (abc)	Final State	Output code word $(v_1v_2v_3)$
00	100	000	100
01	100	010	000
10	100	100	110
11	100	110	010
00	101	000	110
01	101	010	010
10	101	100	100
11	101	110	000
00	110	001	101
01	110	011	001
10	110	101	111
11	110	111	011
00	111	001	111
01	111	011	011
10	111	101	101
11	111	111	001

We shall take $E_b = 1$, so that the 8-PSK signal constellation diagram is as shown below, where R is the code rate $= 2/3$.

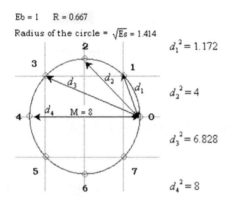

Eb = 1   R = 0.667

Radius of the circle = $\sqrt{Es}$ = 1.414

$d_1^2 = 1.172$

$d_2^2 = 4$

$d_3^2 = 6.828$

$d_4^2 = 8$

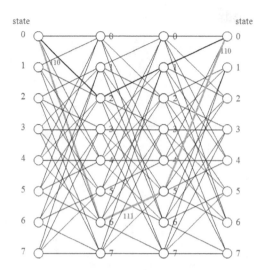

Highlighted path shown corresponds to the minimum free distance path. The code word sequence on this path is 110, 111, 110. Thus, the transmitted signal points are 6, 7, 6 as shown below.

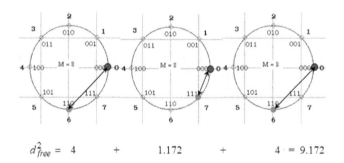

$$d_{free}^2 = 4 \qquad + \qquad 1.172 \qquad + \qquad 4 \ = 9.172$$

Compare this with the path from state 0 —— 2 —— 1 —— 0, which corresponds to the code word sequence 100, 001, 010 and thus the transmitted signal points 4, 1, 2. The distance d_{free}^2 of signal points 4, 1 and 2 from 0, 0, 0 is larger than 9.172 (above).

For QPSK recall that $d_{ref}^2 = 4$. Thus asymptotic coding gain

$$G(dB) = 10\log_{10}\left(\frac{d_{free}^2}{d_{ref}^2}\right) = 10\log_{10}\left(\frac{9.172}{4}\right) = 3.6 \text{ dB} \qquad (8.24)$$

28. Now

$$B_{MPSK} = \frac{2r_b}{R_{code} \log_2 M} \tag{8.25}$$

For no error-control coding,

$$B_{MPSK} = \frac{2r_b}{\log_2 M} \tag{8.26}$$

so that

$$\log_2 M = \frac{2r_b}{B_{MPSK}} \tag{8.27}$$

Hence

$$M = 2^{\frac{2r_b}{B_{MPSK}}} = 2^{\frac{2*40000}{10000}} = 256$$

$$
\begin{aligned}
P_s &= 2Q\left(\sqrt{\frac{2E_s}{N_o}}\sin\left(\frac{\pi}{M}\right)\right) \tag{8.28} \\
&= P_e \log_2 M Q\left(\sqrt{\frac{2E_s}{N_o}}\sin\left(\frac{\pi}{M}\right)\right) \\
&= \frac{P_e \log_2 M}{2} = \frac{0.0001*8}{2} = 0.0004
\end{aligned}
$$

Using the Q-function

$$\sqrt{\frac{2E_s}{N_o}}\sin\left(\frac{\pi}{M}\right) = 3.35, \frac{E_s}{N_o} = \frac{1}{2}\left[\frac{3.38}{\sin\left(\frac{\pi}{M}\right)}\right]^2 \tag{8.29}$$

With no error-control coding, $E_s = E_b \log_2 M$

$$\frac{E_b}{N_o} = \frac{1}{2\log_2 M}\left[\frac{3.35}{\sin\left(\frac{\pi}{M}\right)}\right]^2 \tag{8.30}$$

$$\left(\frac{E_b}{N_o}\right) dB = 10 \log_{10} \left(\frac{1}{2 \log_2 M} \left[\frac{3.35}{\sin\left(\dfrac{\pi}{M}\right)}\right]^2\right) = 36.7 \text{ dB} \qquad (8.31)$$

- ⭐ SIMULATION **P28:** For further details.

29. ⭐ SIMULATION **P29:** For the full solution.

30. ⭐ SIMULATION **P30:** For the full solution.

31. ⭐ SIMULATION **P31:** For the full solution.

32. ⭐ SIMULATION **P32:** For the full solution.

33. ⭐ SIMULATION **P33:** For the full solution.

34. ⭐ SIMULATION **P34:** For the full solution.

35. ⭐ SIMULATION **P35:** For the full solution..

36. (a) Given

$$C = B \log_2 \left(1 + \frac{E_b r_b}{N_o B}\right) \qquad (8.32)$$

with $r_b = C$, we have

$$C = B \log_2 \left(1 + \frac{E_b C}{N_o B}\right) \qquad (8.33)$$

Now let $x = \frac{E_b C}{N_o B}$, so that

$$\frac{C}{B} = \log_2{(1+x)} = x\log_2{(1+x)}^{1/x} = \frac{E_bC}{N_oB}\log_2{(1+x)}^{1/x} \qquad (8.34)$$

Thus

$$1 = \frac{E_b}{N_o}\log_2{(1+x)}^{1/x} \qquad (8.35)$$

As $B \to \infty$, $\frac{C}{B} \to 0$ and thus $x \to 0$. Making use of the identity

$$\lim_{x \to 0}{(1+x)^{1/x}} = e \qquad (8.36)$$

we find

$$1 = \frac{E_b}{N_o}\log_2{e} \qquad (8.37)$$

$$\frac{E_b}{N_o} = \frac{1}{\log_2{e}} \qquad (8.38)$$

$$\begin{aligned}\left(\frac{E_b}{N_o}\right) dB &= 10\log_{10}{\left(\frac{1}{\log_2{e}}\right)} = 10\log_{10}{(\log_2{e})^{-1}} \qquad (8.39)\\ &= -10\log_{10}{(\log_2{e})} = -1.592 \text{ dB}\end{aligned}$$

(b) Let $K = \dfrac{P}{N_o}$. Then

$$C = B\log_2{\left(1+\frac{K}{B}\right)} \qquad (8.40)$$

We find that $x\log_2{\left(1+\dfrac{K}{x}\right)}$ in the limit as $x \to \infty$ is equal to $K\log_2{(e)}$. Thus, a plot of C versus B for a given value of K will show that C approaches the value of $K\log_2{(e)} = 1.44K$. We conclude that the channel capacity always remains finite for a finite signal and noise powers.

● ⭐ SIMULATION **P36:** For further details.

Appendices

Appendix A

Mathematical Tables

A.1 Identities

Trigonometric

$\tan\theta = \frac{\sin\theta}{\cos\theta}$

$\cot\theta = \frac{1}{\tan\theta}$

$\sec\theta = \frac{1}{\cos\theta}$

$\operatorname{cosec}\theta = \frac{1}{\sin\theta}$

$\sin(-\theta) = -\sin\theta$

$\cos(-\theta) = \cos\theta$

$\sin(A \pm B) = \sin A \cos B \pm \cos A \sin B$

$\cos(A \pm B) = \cos A \cos B \mp \sin A \sin B$

$\tan(A \pm B) = \frac{\tan A \pm \tan B}{1 \mp \tan A \tan B}$

$\cos A + \cos B = 2\cos\left(\frac{A+B}{2}\right)\cos\left(\frac{A-B}{2}\right)$

$\cos A - \cos B = -2\sin\left(\frac{A+B}{2}\right)\sin\left(\frac{A-B}{2}\right)$

$\cos\left(\theta \pm \frac{\pi}{2}\right) = \mp\sin\theta$

$\sin\left(\theta \pm \frac{\pi}{2}\right) = \pm\cos\theta$

$\sin\left(90° - \theta\right) = \cos\theta$

$\cos\left(90° - \theta\right) = \sin\theta$

$\cos^2\theta = \frac{1}{2}\left[1 + \cos(2\theta)\right]$

$\sin^2\theta = \frac{1}{2}\left[1 - \cos(2\theta)\right]$

$\cos^3\theta = \frac{1}{4}\left[3\cos\theta + \cos(3\theta)\right]$

$\sin^3\theta = \frac{1}{4}\left[3\sin\theta - \sin(3\theta)\right]$

$\sin^2\theta + \cos^2\theta = 1$

$1 + \cot^2\theta = \operatorname{cosec}^2\theta$

$\cos^2 \theta - \sin^2 \theta = \cos(2\theta)$

$2 \sin \theta \cos \theta = \sin(2\theta)$

$a \cos \theta + b \sin \theta = \sqrt{(a^2 + b^2)} \cos(\theta + \alpha)$, where $\alpha = \tan^{-1}\left(\frac{-b}{a}\right)$ and $a \geq 0$

$a \cos \theta + b \sin \theta = -\sqrt{(a^2 + b^2)} \cos(\theta + \alpha)$, where $\alpha = \tan^{-1}\left(\frac{-b}{a}\right)$ and $a < 0$

$\sin A \sin B = \frac{1}{2}\left[\cos(A - B) - \cos(A + B)\right]$

$\cos A \cos B = \frac{1}{2}\left[\cos(A - B) + \cos(A + B)\right]$

$\sin A \cos B = \frac{1}{2}\left[\sin(A - B) + \sin(A + B)\right]$

Indicies

$x^0 = 1$

$x^p x^q = x^{p+q}$

$\frac{x^p}{x^q} = x^{p-q}$

$\left(x^p\right)^q = x^{pq}$

Logs

$\ln x \equiv \log_e x$

$\log(x^p) = p \log x$

$\log_2 x = \frac{\log_{10}(x)}{\log_{10}(2)}$

$\log A + \log B = \log(AB)$

$\log A - \log B = \log\left(\frac{A}{B}\right)$

Complex

$j = \sqrt{-1}$

$e^{\pm j\theta} = \cos \theta \pm j \sin \theta$

$e^{\pm j\frac{\pi}{2}} = \pm j$

$\cos \theta = \frac{1}{2}\left[e^{j\theta} + e^{-j\theta}\right]$

$\sin \theta = \frac{1}{2j}\left[e^{j\theta} - e^{-j\theta}\right]$

$a + jb = re^{j\theta}$, where $r = \sqrt{(a^2 + b^2)}$ and $\theta = \tan^{-1}\left(\frac{b}{a}\right)$

$r\left(e^{j\theta}\right)^n = r^n e^{jn\theta}$

$\left(r_1 e^{j\theta_1}\right)\left(r_2 e^{j\theta_2}\right) = r_1 r_2 e^{j(\theta_1 + \theta_2)}$

A.2   Series Expansions

Exponential

$e^x = 1 + x + \frac{x^2}{2!} + \frac{x^3}{3!} + \cdots$

Logarithmic

$\log_e (1 + x) = x - \frac{x^2}{2} + \frac{x^3}{3} - \cdots$ where $(|x| < 1)$

Binomial

$(1 + x)^n = 1 + nx + \frac{n(n-1)}{2!} x^2 + \cdots$ where $|nx| < 1$

Taylor

$$f(x) = f(a) + (x - a) \left.\frac{df(x)}{dx}\right|_{x=a} + (x - a)^2 \frac{1}{2!} \left.\frac{d^2 f(x)}{d^2 x}\right|_{x=a} + \cdots + (x - a)^n \frac{1}{n!} \left.\frac{d^n f(x)}{d^n x}\right|_{x=a}$$

- ⭐ SIMULATION **Series:** Experiment with the above series expansions and try various functions for the Taylor series.

A.3   Derivatives and Integrals

$$\frac{d(au)}{dx} = a \frac{d(u)}{dx}$$

$$\frac{d(u+v)}{dx} = \frac{d(u)}{dx} + \frac{d(v)}{dx}$$

$$\frac{d(uv)}{dx} = u \frac{d(v)}{dx} + v \frac{d(u)}{dx}$$

$$\frac{d(x^n)}{dx} = nx^{n-1}$$

$$\frac{d(e^u)}{dx} = e^u \frac{d(u)}{dx}$$

$$\frac{d(e^{ax})}{dx} = ae^{ax}$$

$$\frac{d(\sin u)}{dx} = \cos u \frac{d(u)}{dx}$$

$$\frac{d(\cos u)}{dx} = -\sin u \frac{d(u)}{dx}$$

$$\frac{d(\sin ax)}{dx} = a \cos (ax)$$

$$\frac{d(\cos ax)}{dx} = -a \sin (ax)$$

$$\frac{d(\tan \theta)}{dx} = \sec^2 \theta$$

$$\frac{d(\cot \theta)}{dx} = -\operatorname{cosec}^2 \theta$$

$$\frac{d(\sec \theta)}{dx} = \tan \theta \sec \theta$$

$$\frac{d(\operatorname{cosec} \theta)}{dx} = -\cot \theta \operatorname{cosec} \theta$$

$$\frac{d(\ln ax)}{dx} = \frac{d(\ln x)}{dx} = \frac{1}{x}$$

Indefinite Integrals

$$\int au \, dx = a \int u \, dx$$

$$\int (u + v) \, dx = \int u \, dx + \int v \, dx$$

$$\int x^p \, dx = \frac{x^{p+1}}{p+1} \quad (p \neq -1)$$

$$\int \frac{1}{x} \, dx = \ln |x|$$

$$\int u \, dv = uv - \int v \, du$$

$$\int u \frac{dv}{dx} dx = uv - \int v \frac{du}{dx} dx$$

$$\int \sin(ax) dx = \frac{-1}{a} \cos(ax)$$

$$\int \cos(ax) dx = \frac{1}{a} \sin(ax)$$

$$\int \sin^2(ax) dx = \frac{x}{2} - \frac{\sin(2ax)}{4a}$$

$$\int \cos^2(ax) dx = \frac{x}{2} + \frac{\sin(2ax)}{4a}$$

$$\int x \sin(ax) dx = \frac{\sin(ax) - ax \cos(ax)}{a^2}$$

$$\int x \cos(ax) dx = \frac{\cos(ax) + ax \sin(ax)}{a^2}$$

$$\int x^2 \sin(ax) dx = \frac{2ax \sin(ax) + 2 \cos(ax) - (ax)^2 \cos(ax)}{a^3}$$

$$\int x^2 \cos(ax) dx = \frac{2ax \cos(ax) - 2 \sin(ax) + (ax)^2 \sin(ax)}{a^3}$$

$$\int e^{ax} dx = \frac{e^{ax}}{a}$$

$$\int x e^{ax} dx = \frac{e^{ax}}{a^2} (ax - 1)$$

$$\int x^n e^{ax} dx = \frac{x^n e^{ax}}{a} - \frac{n}{a} \int x^{n-1} e^{ax} dx$$

$$\int x e^{ax^2} dx = \frac{1}{2a} e^{ax^2}$$

$$\int e^{ax} \sin(bx) \, dx = \frac{e^{ax}}{a^2+b^2} \left[a \sin(bx) - b \cos(bx) \right]$$

$$\int e^{ax} \cos(bx) \, dx = \frac{e^{ax}}{a^2+b^2} \left[a \cos(bx) + b \sin(bx) \right]$$

$$\int \frac{1}{(a^2+b^2x^2)} \, dx = \frac{1}{ab} \tan^{-1}\left(\frac{bx}{a}\right)$$

$$\int \frac{x^2}{(a^2+b^2x^2)} \, dx = \frac{x}{b^2} - \frac{a}{b^3} \tan^{-1}\left(\frac{bx}{a}\right)$$

$$\int \frac{x}{(a^2+x^2)} \, dx = \frac{1}{2} \ln(x^2 + a^2)$$

Definite Integrals

$$\int_0^\infty \frac{x \sin(ax)}{b^2+x^2} \, dx = \frac{\pi}{2} e^{(-ab)} \quad \text{where } a > 0 \text{ and } b > 0$$

$$\int_0^\infty \frac{\cos(ax)}{b^2+x^2} \, dx = \frac{\pi}{2b} e^{(-ab)} \quad \text{where } a > 0 \text{ and } b > 0$$

$$\int_0^\infty \text{sinc} \, dx = \int_0^\infty \text{sinc}^2 \, dx = \frac{1}{2}$$

$$\int_0^\infty e^{-ax^2} \, dx = \frac{1}{2} \sqrt{\frac{\pi}{a}} \quad \text{for } a > 0$$

$$\int_0^\infty x^2 e^{-ax^2} \, dx = \frac{1}{4a} \sqrt{\frac{\pi}{a}} \quad \text{for } a > 0$$

$$\int_0^\infty x^n e^{-ax} \, dx = \frac{n!}{a^{n+1}}$$

- ⭐ **SIMULATION Integrals:** Experiment with examples.

CPSIA information can be obtained
at www.ICGtesting.com
Printed in the USA
BVHW092350100720
583349BV00005B/314